T0207151

Practical MATLAB Deep Learning

A Projects-Based Approach

Second Edition

Michael Paluszek

Stephanie Thomas

Eric Ham

Apress®

Practical MATLAB Deep Learning: A Projects-Based Approach

Michael Paluszek
Plainsboro, NJ
USA

Stephanie Thomas
Plainsboro, NJ
USA

Eric Ham
Princeton, NJ
USA

ISBN-13 (pbk): 978-1-4842-7911-3
https://doi.org/10.1007/978-1-4842-7912-0

ISBN-13 (electronic): 978-1-4842-7912-0

Managing Director, Apress Media LLC: Welmoed Spahr
Acquisitions Editor: Steve Anglin
Development Editor: Matthew Moodie
Coordinating Editor: MarkPowers

Cover designed by eStudioCalamar

Cover image by Christopher Burns on Unsplash (www.unsplash.com)

Distributed to the book trade worldwide by Springer Science+Business Media New York, 233 Spring Street, 6th Floor, New York, NY 10013. Phone 1-800-SPRINGER, fax (201) 348-4505, e-mail orders-ny@springer-sbm.com, or visit www.springeronline.com. Apress Media, LLC is a California LLC and the sole member (owner) is Springer Science + Business Media Finance Inc (SSBM Finance Inc). SSBM Finance Inc is a **Delaware** corporation.

For information on translations, please e-mail booktranslations@springernature.com; for reprint, paperback, or audio rights, please e-mail bookpermissions@springernature.com.

Apress titles may be purchased in bulk for academic, corporate, or promotional use. eBook versions and licenses are also available for most titles. For more information, reference our Print and eBook Bulk Sales web page at http://www.apress.com/bulk-sales.

Any source code or other supplementary material referenced by the author in this book is available to readers on GitHub. For more detailed information, please visit http://www.apress.com/source-code.

Contents

About the Authors

Michael Paluszek is President of Princeton Satellite Systems, Inc. (PSS) in Plainsboro, New Jersey. Mr. Paluszek founded PSS in 1992 to provide aerospace consulting services. He used MATLAB to develop the control system and simulations for the IndoStar-1 geosynchronous communications satellite. This led to the launch of Princeton Satellite Systems' first commercial MATLAB toolbox, the Spacecraft Control Toolbox, in 1995. Since then, he has developed toolboxes and software packages for aircraft, submarines, robotics, and nuclear fusion propulsion, resulting in Princeton Satellite Systems' current extensive product line. He is working with the Princeton Plasma Physics Laboratory on a compact nuclear fusion reactor for energy generation and space propulsion.

Before founding PSS, Mr. Paluszek was an engineer at GE Astro Space in East Windsor, NJ. At GE, he designed the Global Geospace Sciences Polar despun platform control system and led the design of the GPS IIR attitude control system, the Inmarsat-3 attitude control system, and the Mars Observer delta-V control system, leveraging MATLAB for control design. Mr. Paluszek also worked on the attitude determination system for the DMSP meteorological satellites. Mr. Paluszek flew communication satellites on over 12 satellite launches, including the GSTAR III recovery, the first transfer of a satellite to an operational orbit using electric thrusters. At Draper Laboratory, Mr. Paluszek worked on the Space Shuttle, early space station, and submarine navigation. His space station work included designing Control Moment Gyro–based control systems for attitude control.

Mr. Paluszek received his bachelor's degree in Electrical Engineering and master's and engineer's degrees in Aeronautics and Astronautics from the Massachusetts Institute of Technology. He is the author of numerous papers and has over a dozen US patents. Mr. Paluszek is the coauthor of *MATLAB Recipes*, *MATLAB Machine Learning*, and *MATLAB Machine Learning Recipes: A Problem-Solution Approach*, all published by Apress.

Stephanie Thomas is Vice President of Princeton Satellite Systems, Inc. in Plainsboro, New Jersey. She received her bachelor's and master's degrees in Aeronautics and Astronautics from the Massachusetts Institute of Technology in 1999 and 2001. Ms. Thomas was introduced to the PSS Spacecraft Control Toolbox for MATLAB during a summer internship in 1996 and has been using MATLAB for aerospace analysis ever since. In her 20 years of MATLAB experience, she has developed many software tools including the Solar Sail Module for the Spacecraft Control Toolbox; a proximity satellite operations toolbox for the Air Force; collision monitoring Simulink blocks for the Prisma satellite mission; and launch vehicle analysis tools in MATLAB and Java. She has developed novel methods for space situation assessment such as a numeric approach to assessing the general rendezvous problem between any two satellites implemented in both MATLAB and C++. Ms. Thomas has contributed to PSS' *Spacecraft Attitude and Orbit Control* textbook, featuring examples using the Spacecraft Control Toolbox, and written many software users' guides. She has conducted SCT training for engineers from diverse locales such as Australia, Canada, Brazil, and Thailand and has performed MATLAB consulting for NASA, the Air Force, and the European Space Agency. Ms. Thomas is the coauthor of *MATLAB Recipes, MATLAB Machine Learning*, and *MATLAB Machine Learning Recipes: A Problem-Solution Approach*, published by Apress. In 2016, Ms. Thomas was named a NASA NIAC Fellow for the project "Fusion-Enabled Pluto Orbiter and Lander."

Eric Ham is an Electrical Engineer and Computer Scientist at Princeton Satellite Systems in Plainsboro, New Jersey. He has a BS in Electrical Engineering with certificates in Applications of Computing and Robotics and Intelligent Systems from Princeton University, 2019. At PSS, Mr. Ham is working on developing neural networks for terrain relative navigation for a lunar lander under a NASA contract. He is simultaneously working as a research specialist with the Hasson Lab at Princeton University's Princeton Neuroscience Institute. He is involved in the design and testing of temporal convolutional neural networks (TCNs) to model semantic processing in the brain. He developed a pipeline for automatic transcription of audio data using Google's speech-to-text API. He assisted in the development of a method for sequence prediction that was inspired by Viterbi's algorithm.

His undergraduate research was on implementing an SDRAM for a novel neuro-computing chip. Mr. Ham did a summer internship at Princeton University in 2018, in which he worked on a novel path selection algorithm to improve the security of the Tor onion router. He worked at the Princeton Plasma Physics Laboratory in 2017 on a high-efficiency Class-E RF amplifier for nuclear fusion plasma heating.

About the Technical Reviewer

 **Dr. Joseph Mueller** specializes in control systems and trajectory optimization. For his doctoral thesis, he developed optimal ascent trajectories for stratospheric airships. His active research interests include robust optimal control, adaptive control, applied optimization and planning for decision support systems, and intelligent systems to enable autonomous operations of robotic vehicles. Prior to joining SIFT in early 2014, Dr. Joseph worked at Princeton Satellite Systems for 13 years. In that time, he served as the principal investigator for eight Small Business Innovation Research contracts for NASA, Air Force, Navy, and MDA. He has developed algorithms for optimal guidance and control of both formation flying spacecraft and high-altitude airships and developed a course of action planning tool for DoD communication satellites. In support of a research study for NASA Goddard Space Flight Center in 2005, Dr. Joseph developed the Formation Flying Toolbox for MATLAB, a commercial product that is now used at NASA, ESA, and several universities and aerospace companies around the world. In 2006, he developed the safe orbit guidance mode algorithms and software for the Swedish Prisma mission, which has successfully flown a two-spacecraft formation flying mission since its launch in 2010. Dr. Joseph also serves as an adjunct professor in the Aerospace Engineering and Mechanics Department at the University of Minnesota, Twin Cities campus.

Acknowledgments

Thanks to Shannen Prindle for helping with the Chapter 7 experiment and doing all of the photography for Chapter 7. Shannen is a Princeton University student who worked as an intern at Princeton Satellite Systems in the summer of 2019. We would also like to thank Dr. Charles Swanson for reviewing Chapter 6 on Tokamak control. Thanks to Kestras Subacius of the MathWorks for tech support on the Bluetooth device. We would also like to thank Matt Halpin for reading the book from front to back. We would like to thank Zaid Zada for his contributions to the chapter on generative deep learning. In particular, we would like to thank Julia Hoerner of the MathWorks for her detailed review of the entire book. She made many excellent suggestions! Thanks also to Dr. Christopher Galea for his help on the Tokamak chapter. We would also like to thank Sam Lehman, Sidhant Shenoy, Emmanouil Tzorako for their support.

We would like to thank dancers Shaye Firer, Emily Parker, 田中棱子(Ryoko Tanaka), and Matanya Solomon for being our experimental subjects in this book. We would also like to thank the American Repertory Ballet and Executive Director Julie Hench for hosting our Chapter 7 experiment.

Preface to the Second Edition

Practical MATLAB Deep Learning, Second Edition, is an extension of the first edition of this book. We have added three new chapters. One shows how deep learning can be applied to the problem of processing static Earth sensor data in low Earth orbit. Many new satellites use Earth sensors. This work shows how you can process data and also use deep learning to evaluate sensors.

The second new chapter is on generative deep learning. This shows how neural networks can be used to generate new data. When a neural network can recognize objects, it has, in its neurons, a model of the subject that allows it to recognize objects it has not seen before. Given this information, a neural network can also create new data. This chapter shows you how. This chapter lets you create a generative deep learning network that generates music.

The final new chapter is on reinforcement learning. Reinforcement learning is a machine learning approach in which an intelligent agent learns to take actions to maximize a reward. We will apply this to the design of a Titan landing control system. Reinforcement learning is a tool to approximate solutions that could have been obtained by dynamic programming, but whose exact solutions are computationally intractable. In this chapter, we derive a model for a Titan lander. We use optimization to come up with a trajectory and then show how reinforcement learning can achieve similar results.

This book was written using several different revisions to MATLAB and its toolboxes. When you replicate the demonstrations, you may notice that some of the GUIs are different. This should not pose a problem with the code that is supplied with the book. This code was tested with R2022a.

CHAPTER 1

■ ■ ■

What Is Deep Learning?

1.1 Deep Learning

Deep learning is a subset of machine learning which is itself a subset of artificial intelligence and statistics. Artificial intelligence research began shortly after World War II [35]. Early work was based on the knowledge of the structure of the brain, propositional logic, and Turing's theory of computation. Warren McCulloch and Walter Pitts created a mathematical formulation for neural networks based on threshold logic. This allowed neural network research to split into two approaches: one centered on biological processes in the brain and the other on the application of neural networks to artificial intelligence. It was demonstrated that any function could be implemented through a set of such neurons and that a neural net could learn to recognize patterns. In 1948, Norbert Wiener's book *Cybernetics* was published which described concepts in control, communications, and statistical signal processing. The next major step in neural networks was Donald Hebb's book in 1949, *The Organization of Behavior*, connecting connectivity with learning in the brain. His book became a source of learning and adaptive systems. Marvin Minsky and Dean Edmonds built the first neural computer at Harvard in 1950.

The first computer programs, and the vast majority now, have knowledge built into the code by the programmer. The programmer may make use of vast databases. For example, a model of an aircraft may use multidimensional tables of aerodynamic coefficients. The resulting software, therefore, knows a lot about aircraft, and running simulations of the models may present surprises to the programmer and the users since they may not fully understand the simulation, or may have entered erroneous inputs. Nonetheless, the programmatic relationships between data and algorithms are predetermined by the code.

In machine learning, the relationships between the data are formed by the learning system. Data is input along with the results related to the data. This is the system training. The machine learning system relates the data to the results and comes up with rules that become part of the system. When new data is introduced, it can come up with new results that were not part of the training set.

Deep learning refers to neural networks with more than one layer of neurons. The name "deep learning" implies something more profound, and in the popular literature, it is taken to imply that the learning system is a "deep thinker." Figure 1.1 shows a single-layer and

M. Paluszek et al., *Practical MATLAB Deep Learning*,
https://doi.org/10.1007/978-1-4842-7912-0_1

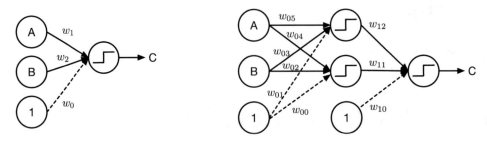

Figure 1.1: *Two neural networks. The one on the right is a deep learning network.*

multi-layer network. It turns out that multi-layer networks can learn things that single-layer networks cannot. The elements of a network are nodes, where weighted signals are combined and biases added. In a single layer, the inputs are multiplied by weights and then added together at the end, after passing through a threshold function. In a multi-layer or "deep learning" network, the inputs are combined in the second layer before being output. There are more weights and the added connections allow the network to learn and solve more complex problems.

There are many types of machine learning. Any computer algorithm that can adapt based on inputs from the environment is a learning system. Here is a partial list:

1. Neural nets (deep learning or otherwise)

2. Support-vector machines

3. Adaptive control

4. System identification

5. Parameter identification (may be the same as the previous one)

6. Adaptive expert systems

7. Control algorithms (a proportional integral derivative control stores information about constant inputs in its integrator)

Some systems use a predefined algorithm and learn by fitting the parameters of the algorithm. Others create a model entirely from data. Deep learning systems are usually in the latter category. We'll next give a brief history of deep learning and then move on to two examples.

1.2 History of Deep Learning

Minsky wrote the book *Perceptrons* with Seymour Papert in 1969, which was an early analysis of artificial neural networks. The book contributed to the movement toward symbolic processing in AI. The book noted that single-layer neurons could not implement some logical functions such as exclusive or (XOR) and implied that multi-layer networks would have the same issue. It was later found that three-layer networks could implement such functions. We give the XOR solution in this book.

Multi-layer neural networks were discovered in the 1960s but not studied until the 1980s. In the 1970s, self-organizing maps using competitive learning were introduced [15]. A resurgence in neural networks happened in the 1980s. Knowledge-based, or "expert," systems were also introduced in the 1980s. From Jackson [18]

> *An expert system is a computer program that represents and reasons with knowledge of some specialized subject to solve problems or give advice.*
>
> —Peter Jackson, *Introduction to Expert Systems*

Backpropagation for neural networks, a learning method using gradient descent, was reinvented in the 1980s leading to renewed progress in this field. Studies began with both human neural networks (i.e., the human brain) and the creation of algorithms for effective computational neural networks. This eventually led to deep learning networks in machine learning applications.

Advances were made in the 1980s as AI researchers began to apply rigorous mathematical and statistical analysis to develop algorithms. Hidden Markov Models were applied to speech. A Hidden Markov Model is a model with unobserved (i.e., hidden) states. Combined with massive databases, they have resulted in vastly more robust speech recognition. Machine translation has also improved. Data mining, the first form of machine learning as it is known today, was developed.

In the early 1990s, Vladimir Vapnik and coworkers invented a computationally powerful class of supervised learning networks known as support-vector machines (SVM). These networks could solve problems of pattern recognition, regression, and other machine learning problems.

There has been an explosion in deep learning in the past few years. New tools have been developed that make deep learning easier to implement. TensorFlow is available from Amazon Web Services (AWS). It makes it easy to deploy deep learning on the cloud. It includes powerful visualization tools. TensorFlow allows you to deploy deep learning on machines that are only intermittently connected to the Web. IBM Watson is another. It allows you to use TensorFlow, Keras, PyTorch, Caffe, and other frameworks. Keras is a popular deep learning framework that can be used in Python. All of these frameworks have allowed deep learning to be deployed just about everywhere.

In this book, we will present MATLAB-based deep learning tools. These powerful tools let you create deep learning systems to solve many different problems. In our book, we will apply MATLAB deep learning to a wide range of problems ranging from nuclear fusion to classical ballet.

Before getting into our examples, we will give some fundamentals on neural nets. We will first give background on neurons and how an artificial neuron represents a real neuron. We will then design a daylight detector. We will follow this with the famous XOR problem that stopped neural net development for some time. Finally, we will discuss the examples in this book.

1.3   Neural Nets

Neural networks, or neural nets, are a popular way of implementing machine "intelligence." The idea is that they behave like the neurons in a brain. In this section, we will explore how neural nets work, starting with the most fundamental idea with a single neuron and working our way up to a multi-layer neural net. Our example for this will be a pendulum. We will show how a neural net can be used to solve the prediction problem. This is one of the two uses of a neural net, prediction and classification. We'll start with a simple classification example.

Let's first look at a single neuron with two inputs. This is shown in Figure 1.2. This neuron has inputs x_1 and x_2, a bias b, weights w_1 and w_2, and a single output z. The activation function σ takes the weighted input and produces the output. In this diagram, we explicitly add icons for the multiplication and addition steps within the neuron, but in typical neural net diagrams such as Figure 1.1, they are omitted.

$$z = \sigma(y) = \sigma(w_1 x_1 + w_2 x_2 + b) \tag{1.1}$$

Let's compare this with a real neuron as shown in Figure 1.3. A real neuron has multiple inputs via the dendrites. Some of these branches mean that multiple inputs can connect to the cell body through the same dendrite. The output is via the axon. Each neuron has one output. The axon connects to a dendrite through the synapse.

There are numerous commonly used activation functions. We show three:

$$\sigma(y) = \tanh(y) \tag{1.2}$$

$$\sigma(y) = \frac{2}{1 - e^{-y}} - 1 \tag{1.3}$$

$$\sigma(y) = y \tag{1.4}$$

The exponential one is normalized and offset from zero so it ranges from -1 to 1. The last one, which simply passes through the value of y, is called the *linear* activation function. The following code in the script `OneNeuron.m` computes and plots these three activation functions for an input q. Figure 1.4 shows the three activation functions on one plot.

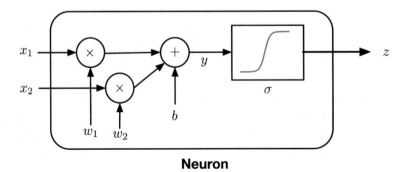

Neuron

Figure 1.2: A two-input neuron.

OneNeuron.m

```
1   %% Single neuron demonstration.
7   %% Look at the activation functions
8   y        = linspace(-4,4);
9   z1       = tanh(y);
10  z2       = 2./(1+exp(-y)) - 1;
11
12  PlotSet(y,[z1;z2;y],'x label','Input', 'y label',...
13     'Output', 'figure title','Activation Functions','plot title', '
           Activation Functions',...
14     'plot set',{[1 2 3]},'legend',{{'Tanh','Exp','Linear'}});
```

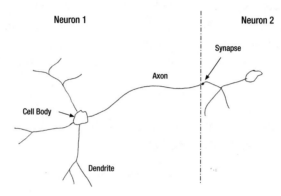

Figure 1.3: *A neuron connected to a second neuron. A real neuron can have 10,000 inputs!*

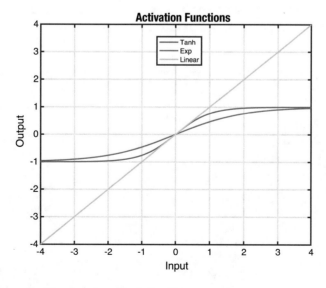

Figure 1.4: *The three activation functions from* OneNeuron.

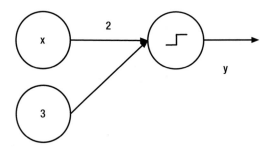

Figure 1.5: *A one-input neural net. The weight w is 2 and the bias b is 3.*

Activation functions that saturate, or reach a value of input after which the output is constant or changes very slowly, model a biological neuron that has a maximum firing rate. These particular functions also have good numerical properties that are helpful in learning.

Let's look at a single input neural net shown in Figure 1.5. This neuron is

$$z = \sigma(2x + 3) \tag{1.5}$$

where the weight w on the single input x is 2 and the bias b is 3. If the activation function is linear, the neuron is just a linear function of x:

$$z = y = 2x + 3 \tag{1.6}$$

Neural nets do make use of linear activation functions, often in the output layer. It is the nonlinear activation functions that give neural nets their unique capabilities.

Let's look at the output with the preceding activation functions plus the threshold function from the script `LinearNeuron.m`. The results are in Figure 1.6.

LinearNeuron.m

```
1   %% Linear neuron demo
6   x        = linspace(-4,2,1000);
7   y        = 2*x + 3;
8   z1       = tanh(y);
9   z2       = 2./(1+exp(-y)) - 1;
10  z3       = zeros(1,length(x));
11
12  % Apply a threshold
13  k        = y >=0;
14  z3(k)    = 1;
15
16  PlotSet(x,[z1;z2;z3;y],'x label','x', 'y label',...
17     'y', 'figure title','Linear Neuron','plot title', 'Linear Neuron',...
18     'plot set',{[1 2 3 4]},'legend',{{'Tanh','Exp','Threshold','Linear'
             }});
```

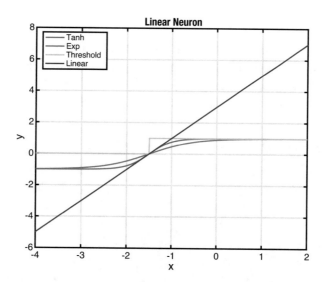

Figure 1.6: *The "linear" neuron compared to other activation functions from* `LinearNeuron`.

The `tanh` and `exp` are very similar. They put bounds on the output. Within the range $-3 \leq x < 1$, they return the function of the input. Outside those bounds, they return the sign of the input, that is, they saturate. The threshold function returns 0 if the value is less than 0 and 1 if it is greater than -1.5. The threshold is saying the output is only important, thus *activated*, if the input exceeds a given value. The other nonlinear activation functions are saying that we care about the value of the linear equation only within the bounds. The nonlinear functions (but not steps) make it easier for the learning algorithms since the functions have derivatives. The binary step has a discontinuity at an input of zero so that its derivative is infinite at that point. Aside from the linear function (which is usually used on output neurons), the neurons are just telling us that the sign of the linear equation is all we care about. The activation function is what makes a neuron a neuron.

We now show two brief examples of neural nets: first, a daylight detector, and second, the exclusive or problem.

1.3.1 Daylight Detector

Problem

We want to use a simple neural net to detect daylight. This will provide an example of using a neural net for classification.

Solution

Historically, the first neuron was the perceptron. This is a neuron with an activation function that is a threshold. Its output is either 0 or 1. This is not useful for many real-world problems. However, it is well suited for simple classification problems. We will use a single perceptron in this example.

How It Works

Suppose our input is a light level measured by a photocell. If you weight the input so that 1 is the value defining the brightness level at noon, you get a sunny day detector.

This is shown in the script, `SunnyDay.m`. The solar flux is modeled using cosine and scaled so that it is 1 at noon. Any value greater than 0 is daylight.

SunnyDay.m

```
8   %% The data
9   t = linspace(0,24);        % time, in hours
10  d = zeros(1,length(t));
11  s = cos((2*pi/24)*(t-12)); % solar flux model
12
13  %% The activation function
14  % The nonlinear activation function which is a threshold detector
15  j     = s < 0;
16  s(j)  = 0;
17  j     = s > 0;
18  d(j)  = 1;
19
20  %% Plot the results
21  PlotSet(t,[s;d],'x label','Hour', 'y label',...
22      {'Solar Flux', 'Day/Night'}, 'figure title','Daylight Detector',...
23      'plot title', {'Flux Model','Perceptron Output'});
24  set([subplot(2,1,1) subplot(2,1,2)],'xlim',[0 24],'xtick',[0 6 12 18
        24]);
```

Figure 1.7 shows the detector results. The `set(gca,...)` code sets the x-axis ticks to end at exactly 24 hours. This is a really trivial example but does show how classification works.

If we had multiple neurons with thresholds set to detect sunlight levels within bands of solar flux, we would have a neural net sun clock.

1.3.2 XOR Neural Net

Problem

We want to implement the exclusive or (XOR) problem with a neural network.

Solution

The XOR problem impeded the development of neural networks for a long time before "deep learning" was developed. Look at Figure 1.8. The table on the left gives all possible inputs A and B and the desired outputs C. "Exclusive or" just means that if the inputs A and B are different, the output C is 1. The figure shows a single-layer network and a multi-layer network, as in Figure 1.1, but with the weights labeled as they will be in the code. You can implement this in MATLAB easily, in just seven lines:

```
>> a = 1;
>> b = 0;
>> if( a == b )
>>    c = 1
>> else
>>    c = 0
>> end

c =
      0
```

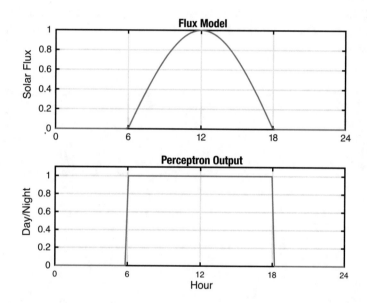

Figure 1.7: *The daylight detector. The top plot shows the input data, and the bottom plot shows the perceptron output detecting daylight.*

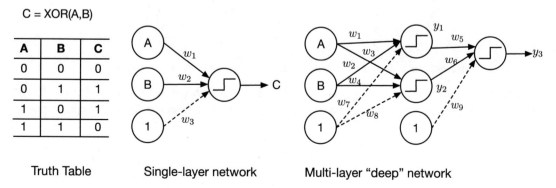

Figure 1.8: *Exclusive or (XOR) truth table and possible solution networks.*

9

This type of logic was embodied in medium-scale integrated circuits in the early days of digital systems and vacuum tube–based computers even earlier than that. Try as you might, you cannot pick two weights and a bias on the single-layer network to reproduce the XOR. Minsky created proof that it was impossible.

The second neural net, the deep neural net, can reproduce the XOR. We will implement and train this network.

How It Works

What we will do is explicitly write out the backpropagation algorithm that trains the neural net from the four training sets given in Figure 1.8, that is, (0,0), (1,0), (0,1), or (1,1). We'll write it in the script XORDemo.m. The point is to show you explicitly how backpropagation works. We will use the tanh as the activation function in this example. The XOR function is given in XOR.m as follows:

XOR.m

```
1   %% XOR Implement an 'Exclusive Or' neural net
5   %   c = XOR(a,b,w)
6   %
7   %% Description
8   % Implements an XOR function in a neural net. It accepts vector inputs.
9   %
10  %% Inputs
11  %   a   (1,:)   Input 1
12  %   b   (1,:)   Input 2
13  %   w   (9,1)   Weights and biases
14  %% Outputs
15  %   c   (1,:)   Output
16  %
17  function [y3,y1,y2] = XOR(a,b,w)
18
19  if( nargin < 1 )
20    Demo
21    return
22  end
23
24  y1 = tanh(w(1)*a  + w(2)*b  + w(7));
25  y2 = tanh(w(3)*a  + w(4)*b  + w(8));
26  y3 = w(5)*y1 + w(6)*y2 + w(9);
27  c  = y3;
```

There are three neurons, y_1, y_2, and y_3. The activation function for the hidden layer with neurons y_1 and y_2 is the hyperbolic tangent. The activation function for the output layer y_3

is linear. In addition to the weights depicted in Figure 1.8, each neuron also has a bias input, numbered w_7, w_8, and w_9:

$$y_1 \;=\; \tanh(w_1 a + w_2 b + w_7) \tag{1.7}$$

$$y_2 \;=\; \tanh(w_3 a + w_4 b + w_8) \tag{1.8}$$

$$y_3 \;=\; w_5 y_1 + w_6 y_2 + w_9 \tag{1.9}$$

Now we will derive the backpropagation routine. The hyperbolic activation function is

$$f(z) = \tanh(z) \tag{1.10}$$

Its derivative is

$$\frac{df(z)}{dz} = 1 - f^2(z) \tag{1.11}$$

In this derivation, we are going to use the chain rule. Assume that F is a function of y which is a function of x. Then

$$\frac{dF(y(x))}{dx} = \frac{dF}{dy}\frac{dy}{dx} \tag{1.12}$$

The error is the square of the difference between the desired output and the output. This is known as a quadratic error. It is easy to use because the derivative is simple and the error is always positive, making the lowest error the one closest to zero.

$$E = \frac{1}{2}\left(c - y_3\right)^2 \tag{1.13}$$

The derivative of the error for w_j for the output node

$$\frac{\partial E}{\partial w_j} = (y_3 - c)\frac{\partial y_3}{\partial w_j} \tag{1.14}$$

For the hidden nodes, it is

$$\frac{\partial E}{\partial w_j} = \psi_3 \frac{\partial n_3}{\partial w_j} \tag{1.15}$$

Expanding for all the weights

$$\frac{\partial E}{\partial w_1} \;=\; \psi_3 \psi_1 a \tag{1.16}$$

$$\frac{\partial E}{\partial w_2} \;=\; \psi_3 \psi_1 b \tag{1.17}$$

$$\frac{\partial E}{\partial w_3} \;=\; \psi_3 \psi_2 a \tag{1.18}$$

$$\frac{\partial E}{\partial w_4} \;=\; \psi_3 \psi_2 b \tag{1.19}$$

$$\frac{\partial E}{\partial w_5} \;=\; \psi_3 y_1 \tag{1.20}$$

$$\frac{\partial E}{\partial w_6} \;=\; \psi_3 y_2 \tag{1.21}$$

$$\frac{\partial E}{\partial w_7} \;=\; \psi_3 \psi_1 \tag{1.22}$$

$$\frac{\partial E}{\partial w_8} \;=\; \psi_3 \psi_2 \tag{1.23}$$

$$\frac{\partial E}{\partial w_9} \;=\; \psi_3 \tag{1.24}$$

where

$$\psi_1 \;=\; 1 - f^2(n_1) \tag{1.25}$$

$$\psi_2 \;=\; 1 - f^2(n_2) \tag{1.26}$$

$$\psi_3 \;=\; y_3 - c \tag{1.27}$$

$$n_1 \;=\; w_1 a + w_2 b + w_7 \tag{1.28}$$

$$n_2 \;=\; w_3 a + w_4 b + w_8 \tag{1.29}$$

$$n_3 \;=\; w_5 y_1 + w_6 y_2 + w_9 \tag{1.30}$$

You can see from the derivation how this could be made recursive and applied to any number of outputs or layers. Our weight adjustment at each step will be

$$\Delta w_j = -\eta \frac{\partial E}{\partial w_j} \tag{1.31}$$

where η is the update gain. It should be a small number. We only have four sets of inputs. We will apply them multiple times to get the XOR weights.

Our backpropagation trainer needs to find the nine elements of w. The training function XORTraining.m is as follows:

XORTraining.m

```
 1  %% XORTRAINING Implements an XOR training function.
 9  %% Inputs
10  %   a     (1,4)  Input 1
11  %   b     (1,4)  Input 2
12  %   c     (1,4)  Output
13  %   w     (9,1)  Weights and biases
14  %   n     (1,1)  Number of iterations through all 4 inputs
15  %   eta   (1,1)  Training weight
16  %
17  %% Outputs
18  %   w     (9,1)  Weights and biases
19  %% See also
20  %  XOR
21
```

```
22  function w = XORTraining(a,b,c,w,n,eta)
23
24  if( nargin < 1 )
25    Demo;
26    return
27  end
28
29  e          = zeros(4,1);
30  y3         = XOR(a,b,w);
31  e(:,1)     = y3 - c;
32  wP         = zeros(10,n+1); % For plotting the weights
33  for k = 1:n
34    wP(:,k) = [w;mean(abs(e))];
35    for j = 1:4
36      [y3,y1,y2]  = XOR(a(j),b(j),w);
37      psi1        = 1 - y1^2;
38      psi2        = 1 - y2^2;
39      e(j)        = y3 - c(j);
40      psi3        = e(j); % Linear activation function
41      dW          = psi3*[psi1*a(j);psi1*b(j);psi2*a(j);psi2*b(j);y1;y2;
                     psi1;psi2;1];
42      w           = w - eta*dW;
43    end
44  end
45  wP(:,k+1) = [w;mean(abs(e))];
46
47  % For legend entries
48  wName = string;
49  for k = 1:length(w)
50    wName(k) = sprintf('W_%d',k);
51  end
52  leg{1} = wName;
53  leg{2} = '';
54
55  PlotSet(0:n,wP,'x label','step','y label',{'Weight' 'Error'},...
56    'figure title','Learning','legend',leg,'plot set',{1:9 10});
```

The first two arguments to `PlotSet` are the data and are the minimum required. The remainder is parameter pairs. The `leg` value has legends for the two plots, as defined by `'plot set'`. The first plot uses the first nine data points, in this case, the weights. The second plot uses the last data point, the mean of the error. `leg` is a cell array with two strings or string arrays. The `'plot set'` is two arrays in a cell. A plot with only one value will not generate a legend.

The demo script `XORDemo.m` starts with the training data, which is the complete truth data for this simple function, and randomly generated weights. It iterates through the inputs 25,000 times, with a training weight of 0.001.

XORDemo.m

```
5  % Training data - also the truth data
6  a       = [0 1 0 1];
7  b       = [0 0 1 1];
8  c       = [0 1 1 0];
9
10 % First try implementing random weights
11 w0      = [ 0.1892; 0.2482; -0.0495; -0.4162; -0.2710;...
12            0.4133; -0.3476; 0.3258; 0.0383];
13 cR      = XOR(a,b,w0);
14
15 fprintf('\nRandom Weights\n')
16 fprintf('    a     b    c\n');
17 for k = 1:4
18   fprintf('%5.0f %5.0f %5.2f\n',a(k),b(k),cR(k));
19 end
20
21 % Now execute the training
22 w       = XORTraining(a,b,c,w0,25000,0.001);
23 cT      = XOR(a,b,w);
```

The results of the neural network with random weights and biases, as expected, are not good. After training, the neural network reproduces the XOR problem very well, as shown in the following demo output. Now, if you change the initial weights and biases, you may find that you get bad results. This is because the simple gradient method implemented here can fall into local minima from which it can't escape. This is an important point about finding the best answer. There may be many good answers, which are locally optimal, but there will be only one best answer. There is a vast body of research on how to guarantee that a solution is globally optimal.

```
1  >> XORDemo
2
3  Random Weights
4       a     b    c
5       0     0  0.26
6       1     0  0.19
7       0     1  0.03
8       1     1 -0.04
9
10 Weights and Biases
11   Initial    Final
12   0.1892    1.7933
13   0.2482    1.8155
14  -0.0495   -0.8535
15  -0.4162   -0.8591
16  -0.2710    1.3744
17   0.4133    1.4893
18  -0.3476   -0.4974
19   0.3258    1.1124
```

```
20      0.0383 -0.5634
21
22  Trained
23      a      b    c
24      0      0    0.00
25      1      0    1.00
26      0      1    1.00
27      1      1    0.01
```

Figure 1.9 shows the weights and biases converging and also shows the mean output error over all four inputs in the truth table going to zero. If you try other starting weights and biases, this may not be the case. Other solution methods, such as Genetic Algorithms [14], Electromagnetism based [4], and Simulated Annealing [32], are less susceptible to falling into local minima but can be slow. A good overview of optimization specifically for machine learning is given by Bottou [5].

In the next chapter, we will use the MATLAB Deep Learning Toolbox to solve this problem.

You can see how this compares to a set of linear equations. If we remove the activation functions, we get

$$y_3 = w_9 + w_6 w_8 + w_5 w_7 + a(w_1 w_5 + w_3 w_6) + b(w_2 w_5 + 2_r w_6) \tag{1.32}$$

This reduces to just three independent coefficients:

$$y_3 = k_1 + k_2 a + k_3 b \tag{1.33}$$

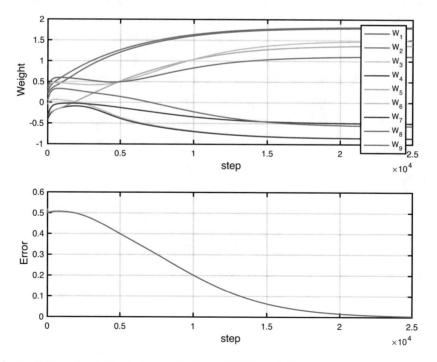

Figure 1.9: *Evolution of weights during exclusive or (XOR) training.*

15

One is a constant, and the other two multiply the inputs. Writing the four possible cases in matrix notation, we get

$$\begin{bmatrix} 0 \\ 1 \\ 1 \\ 0 \end{bmatrix} = k_1 + \begin{bmatrix} 0 & 1 \\ 0 & 0 \\ 1 & 1 \\ 1 & 0 \end{bmatrix} \begin{bmatrix} k_2 \\ k_3 \end{bmatrix} \tag{1.34}$$

We can get close to a working XOR if we choose

$$\begin{bmatrix} k_1 \\ k_2 \\ k_3 \end{bmatrix} = \begin{bmatrix} 1 \\ -1 \\ -1 \end{bmatrix} \tag{1.35}$$

This makes three out of four equations correct. There is no way to make all four correct with just three coefficients. The activation functions separate the coefficients and allow us to reproduce the XOR. This is not surprising because the XOR is not a linear problem.

1.4 Deep Learning and Data

Deep learning systems operate on data but, unlike other systems, have multiple layers that learn the input/output relationships. Data may be organized in many ways. For example, we may want a deep learning system to identify an image. A color image that is 2 pixels by 2 pixels by 3 colors could be represented with random data using `rand`:

```
1   >> x = rand(2,2,3)
2
3   x(:,:,1) =
4
5        0.9572    0.8003
6        0.4854    0.1419
8
9   x(:,:,2) =
10
11       0.4218    0.7922
12       0.9157    0.9595
14
15  x(:,:,3) =
16
17       0.6557    0.8491
18       0.0357    0.9340
```

The array form implies a structure for the data. The same number of points could be organized into a single vector using `reshape`:

```
1   >> reshape(x,12,1)
2
3   ans =
4
5       0.9572
6       0.4854
7       0.8003
8       0.1419
9       0.4218
10      0.9157
11      0.7922
12      0.9595
13      0.6557
14      0.0357
15      0.8491
16      0.9340
```

The numbers are the same, they are just organized differently. Convolutional neural networks, described in the next section, are often used for image structured data. We might also have a vector:

```
1   >> s = rand(2,1)
2
3   s =
4
5       0.6787
6       0.7577
```

for which we wish to learn a temporal or time sequence. In this case, if each column is a time sample, we might have

```
1   >> rand(2,4)
2
3   and =
4
5       0.7431    0.6555    0.7060    0.2769
6       0.3922    0.1712    0.0318    0.0462
```

For example, we might want to look at an ongoing sequence of samples and determine if a set of k samples matches a predetermined sequence. For this simple problem, the neural net would learn the sequence and then be fed sets of four samples to match.

We also need to distinguish what we mean by matching. If all of our numbers are exact, the problem is relatively straightforward. In real systems, measurements are often noisy. In those cases, we want to match with a certain probability. This leads to the concept of statistical neural nets.

1.5 Types of Deep Learning

There are many types of deep learning networks. New types are under development as you read this book. One deep learning researcher joked that you will have the name for an existing deep learning algorithm if you randomly put together four letters.

The following sections briefly describe some of the major types.

1.5.1 Multi-layer Neural Network

Multi-layer neural networks have

1. Input neurons

2. Multiple layers of hidden neurons

3. Output neurons

The different layers may have different activation functions. They may also be functionally different such as being a convolution or pooling layer. In a later chapter, we will introduce the idea of an algorithmic layer.

1.5.2 Convolutional Neural Network (CNN)

A CNN has convolutional layers. It convolves a feature with the input matrix so that the output emphasizes that feature. This effectively finds patterns. For example, you might convolve an L pattern with the incoming data to find corners. The human eye has edge detectors, making the human vision system a convolutional neural network of sorts.

1.5.3 Recurrent Neural Network (RNN)

Recurrent neural networks are a type of recursive neural network. Recurrent neural networks are often used for time-dependent problems. They combine the last time step's data with the data from the hidden or intermediate layer, to represent the current time step. A recurrent neural net has a loop. An input vector at time k is used to create an output which is then passed to the next element of the network. This is done recursively in that each stage is identical to external inputs and inputs from the previous stage. Recurrent neural nets are used in speech recognition, language translation, and many other applications. One can see how a recurrent network would be useful in translation. The meaning of the latter part of an English sentence can be dependent on the beginning. Now, this presents a problem. Suppose we are translating a paragraph. Is the output of the first stage necessarily relevant to the 100th stage? In standard estimation, old data is forgotten using a forgetting factor. In neural networks, we can use Long Short-Term Memory (LSTM) networks that have this feature.

1.5.4 Long Short-Term Memory Network (LSTM)

LSTMs are designed to avoid the dependency on old information. A standard RNN has a repeating structure. An LSTM also has a repeating structure, but each element has four layers. The LSTM layers decide what old information to pass on to the next layer. It may be all, or it may be none. There are many variants of LSTM, but they all include the fundamental ability to forget things.

1.5.5 Recursive Neural Network

This is often confused with recurrent neural networks (RNNs), which are a type of recursive neural network. Recursive neural networks operate on structured data. Any type of structure can be used. RNNs have time as the structuring element. They've been used successfully in language processing as language is structured (as opposed to images that are not).

1.5.6 Temporal Convolutional Machine (TCM)

The TCM is a convolutional architecture designed to learn temporal sequences [24]. TCMs are particularly useful for the statistical modeling of temporal sequences. Statistical modeling is appropriate when incoming data is noisy.

1.5.7 Stacked Autoencoders

A stacked autoencoder is a neural net made up of a series of sparse autoencoders. An autoencoder is a type of neural network that is an unsupervised learning algorithm using backpropagation. Sparsity is a measure of how many neurons are activated, that is, have inputs that cause them to produce an output for a given activation function. The outputs of one layer feed into the next. The number of nodes tends to decrease as you move from input to output.

1.5.8 Extreme Learning Machine (ELM)

ELMs were invented by Guang-Bin Huang [17]. ELMs are a single hidden layer feedforward network. It randomly chooses the weights of the hidden nodes and analytically computes the weights of the output nodes. ELMs provide good performance and learn quickly.

1.5.9 Recursive Deep Learning

Recursive deep learning [39] is a variation and extension of RNNs. The same set of neural node weights is applied recursively over a structured input. That is, not all of the inputs are processed in batches. Recursion is a standard method used in general estimation when data is coming in at different times and you want the best estimate at the current time without having to process all available data at once.

1.5.10 Generative Deep Learning

Generative deep learning allows a neural network to learn patterns [13] and then create completely new material. A generative deep learning network can create articles, paintings, photographs, music, and many other types of material. Generative deep learning is explored in Chapter 14.

1.5.11 Reinforcement Learning

Reinforcement learning allows a neural network to learn to do a task for which there is a reward for performing the task well. Reinforcement learning is explored in Chapter 15.

1.6 Applications of Deep Learning

Deep learning is used in many applications today. Here are a few:

Image recognition – This is arguably the best known and most controversial use of deep learning. A deep learning system is trained with pictures of people. Cameras are distributed everywhere, and images are captured. The system then identifies individual faces and matches them against its trained database. Even with variations in lighting, weather conditions, and clothing, the system can identify the people in the images.

Speech recognition – You hardly ever get a human being on the phone anymore. You are first presented with a robotic listener that can identify what you are saying, at least within the limited context of what it expects. When a human listens to another human, the listener is not just recording the speech, they are guessing what the person is going to say and filling in gaps of garbled words and confusing grammar. Robotic listeners have some of the same abilities. A robotic listener is an embodiment of the "Turing test." Did you ever get one that you thought was a human being? Or for that matter, did you ever reach a human who you thought was a robot?

Handwriting analysis – A long time ago, you would get forms in which you had boxes in which to write numbers and letters. At first, they had to be block capitals! A robotic handwriting system could figure out the letters in those boxes reliably. Years later, though many years ago, the US Post Office introduced zip code reading systems. At first, you had to put the zip code on a specific part of the envelope. That system has evolved so that it can find zip codes anywhere. This made the zip + 4 system valuable and a big productivity boost.

Machine translation – Google translate does a pretty good job considering it can translate almost any language in the world. It is an example of a system with online training. You see that when you type in a phrase and the translation has a checkmark next to it because a human being has indicated that it is correct. Figure 1.10 gives an example. Google harnesses the services of free human translators to improve its product!

Targeting – By targeting, we mean figuring out what you want. This may be a movie, a clothing item, or a book. Deep learning systems collect information on what you like and

Figure 1.10: *Translation from Japanese into English.*

Figure 1.11: *Prediction of your buying patterns.*

decide what you would be most interested in buying. Figure 1.11 gives an example. This is from a couple of years ago. Perhaps, ballet dancers like *Star Wars*!

Other applications include game playing, autonomous driving, medicine, and many others. Just about any human activity can be an application of deep learning.

1.7   Organization of the Book

This book is organized around specific deep learning examples. You can jump into any chapter as they are pretty much independent. We've tried to present a wide range of topics, some of which, hopefully, align with your work or interests. The next chapter gives an overview of MATLAB products for deep learning. Besides the core MATLAB development environment, we only use three of their toolboxes in this book.

Each chapter except for this and the next is organized in the following order:

1. Modeling

2. Building the system

3. Training the system

4. Testing the system

Training and testing are often in the same script. Modeling varies with each chapter. For physical problems, we derive numerical models, usually sets of differential equations, and build simulations of the processes.

The chapters in this book present a range of relatively simple examples to help you learn more about deep learning and its applications. It will also help you learn the limitations of deep learning and areas for future research. All use the MATLAB Deep Learning Toolbox.

1. **What Is Deep Learning?** (this chapter).

2. **MATLAB Machine Learning Toolboxes** – This chapter gives you an introduction to MATLAB machine intelligence toolboxes. We'll be using three of the toolboxes in this book.

3. **Finding Circles with Deep Learning** – This is an elementary example. The system will try to figure out if a figure is a circle. It will be presented with circles, ellipses, and other objects and trained to determine which are circles.

4. **Classifying Movies** – All movie databases try to guess what movies will be of most interest to their viewers to speed movie selection and reduce the number of disgruntled customers. This example creates a movie rating system and attempts to classify movies in the movie database as good or bad.

5. **Algorithmic Deep Learning** – This is an example of fault detection using a detection filter as an element of the deep learning system. It uses a custom deep learning algorithm, the only example that does not use the MATLAB Deep Learning Toolbox.

6. **Tokamak Disruption Detection** – Disruptions are a major problem with a nuclear fusion device known as a Tokamak. Researchers are using neural nets to detect disruptions before they happen so that they can be stopped. In this example, we use a simplified dynamical model to demonstrate deep learning.

7. **Classifying a Pirouette** – This example demonstrates how to use real-time data in a deep learning system. It uses IMU data, via Bluetooth, and camera input. The data is combined to classify a dancer's pirouette (turn). This example will also cover data acquisition and using real-time data as part of a deep learning system.

8. **Completing Sentences** – Writing systems sometimes attempt to predict what word or sentence fragment you are trying to use. We create a database of sentences and try to predict the remainder.

9. **Terrain-Based Navigation** – The first cruise missiles used terrain mapping to reach their targets. This has been largely replaced by GPS. This system will identify where an aircraft is on a map and use past positions to predict future positions.

10. **Stock Prediction** – Who wouldn't want a system that could create portfolios that would beat an index fund? Perhaps a stock prediction system that would find the next Apple or Microsoft! In this example, we create an artificial stock market and train the system to identify the best stocks.

11. **Image Classification** – Training deep learning networks can take weeks. This chapter gives you an example of using a pretrained network.

12. **Orbit Determination** – Orbits can be determined using only angle measurements. This chapter shows `fitnet`, one of the Deep Learning Toolboxes neural network generating functions, can produce estimates of the semi-major axis and eccentricity from angles.

13. **Earth Sensors** – Machine learning can be used to map sensor outputs to roll and pitch estimates.

14. **Generative Modeling of Music** – Machine learning can create new data. In this case a machine learning algorithm generates music.

15. **Reinforcement Learning** – Reinforcement learning can be used in many places where optimization is used. This chapter compares optimization and reinforcement learning for a Titan landing problem.

These are all very different problems. We give a summary of the theory behind each which is hopefully enough for you to understand the problem. There are hundreds of papers on each topic, and even textbooks on these subjects. The references provide more information. There are two broad methods for applying deep learning in MATLAB. One is using `trainNetwork`, and the other is using the various feedforward functions. Table 1.1 summarizes which methods are used in each chapter.

Chapter 11 uses pretrained networks, but these are similar to those produced by `train Network`. Chapter 12 applies four types of network training to the same problem. Chapter 13 shows how neural networks can be used to convert detector measurements into roll and pitch in an Earth sensor. Chapter 14 gives an example of generative deep learning in which a neural

Table 1.1: *Deep learning methods. The specific form of the network is shown. The last column shows the application.*

Chapter	Feedforward	Network Type	Type
2	`feedforwardnet`		Regression
3		Convolutional	Image classification
4	`patternnet`		Classification
5	`feedforwardnet`		Regression
6		Bidirectional LSTM	Classification
7		Bidirectional LSTM	Classification
8		Bidirectional LSTM	Classification
9		Convolutional	Image classification
10		LSTM	Regression
11		**	Image classification
12	`feedforwardnet, fitnet, cascadeforwardnet`	Bidirectional LSTM	Regression

network creates music based on what it has learned. Chapter 15 presents reinforcement learning in the context of a Titan lander.

For each problem, we are creating a world in which to work. For example, in the classifying movies problem, we create a world of movies and viewers based on a particular model that we create. This is akin to the famous "Blocks World" in which a world of colored blocks was created. The artificial intelligence engine could reason and solve problems of stacking blocks within the context of this world. Much like "Blocks World" did not map into general reasoning, we do not claim that our code can be applied directly to real-world problems.

In each chapter, we will present a problem and give code that creates a deep learning network to solve the problem. We will demonstrate the performance of the code but will also show you the limitations and where the code doesn't work as well as we would like. Deep learning is a work in progress, and it is important to understand what works and what doesn't work. We encourage our readers to go beyond the code in the book and see if they can improve on its performance!

We present much of the code in segments. Unless specified, you cannot cut and paste the code into the MATLAB command window and get a result. You should run the demos from the code base that is included with the book. Remember also that you will need the MATLAB Deep Learning Toolbox and Instrument Control Toolbox for Chapter 7. The other chapters only require core MATLAB and the MATLAB Deep Learning Toolbox.

The code in this book was developed using MATLAB 2019a through 2022a on an Apple MacBook Pro under MacOS 10.14.4. The code should work on all other operating systems though processing time may vary. The code can be found at `https://github.com/Apress/practical-matlab-deep-learning-2e`.

CHAPTER 2

■ ■ ■

MATLAB Machine Learning Toolboxes

2.1 Commercial MATLAB Software

2.1.1 MathWorks Products

The MathWorks sells several packages for machine learning. Their toolboxes work directly with MATLAB and Simulink. The MathWorks products provide high-quality algorithms for data analysis along with graphics tools to visualize the data. Visualization tools are a critical part of any machine learning system. They can be used for data acquisition, for example, for image recognition or as part of systems for autonomous control of vehicles, or for diagnosis and debugging during development. All of these packages can be integrated with each other and with other MATLAB functions to produce powerful systems for machine learning. The most applicable toolboxes that we will discuss are listed in the following; we will use only the Deep Learning and the Instrument Control toolboxes in this book:

- **Deep Learning Toolbox**
- **Instrument Control Toolbox**
- Statistics and Machine Learning Toolbox
- Computer Vision Toolbox
- Image Acquisition Toolbox
- Parallel Computing Toolbox
- Text Analytics Toolbox

The breadth of MATLAB products allow you to explore every facet of machine learning and to connect with other areas of data science including controls, estimation, and simulation. There are also many domain-specific toolboxes, such as the Automated Driving Toolbox and Sensor Fusion and Tracking Toolbox that can be used with the learning products.

© Michael Paluszek, Stephanie Thomas, Eric Ham 2022
M. Paluszek et al., *Practical MATLAB Deep Learning*,
https://doi.org/10.1007/978-1-4842-7912-0_2

Deep Learning Toolbox

The Deep Learning Toolbox allows you to design, build, train, and visualize a wide variety of neural networks. You can implement existing, pretrained neural networks available on the Web, such as GoogLeNet, VGG-16, VGG-19, AlexNet, and ResNet-59. GoogLeNet and AlexNet are image classification networks and are discussed in Chapter 11. The Deep Learning Toolbox has extensive capabilities for visualization and debugging of neural networks. The debugging tools are important to ensure that your system is behaving properly and help you to understand what is going on inside your neural network. It includes a number of pretrained models. We will use this toolbox in all of our examples.

Instrument Control Toolbox

The MATLAB Instrument Control Toolbox is designed to directly connect instruments. This simplifies the use of MATLAB with hardware. Examples include oscilloscopes, function generators, and power supplies. The toolbox provides support for TCP/IP, UDP, I2C, SPI, and Bluetooth. With the Instrument Control Toolbox, you can integrate MATLAB directly into your laboratory workflow without the need for writing drivers or creating specialized MEX files. We use the Bluetooth functionality with an IMU in this book.

Statistics and Machine Learning Toolbox

The Statistics and Machine Learning Toolbox provides data analytics methods for gathering trends and patterns from massive amounts of data. These methods do not require a model for analyzing the data. The toolbox functions can be broadly divided into classification tools, regression tools, and clustering tools. Statistics are the foundation for much of deep learning.

Classification methods are used to place data into different categories. For example, data, in the form of an image, might be used to classify an image of an organ as having a tumor. Classification is used for handwriting recognition, credit scoring, and face identification. Classification methods include support-vector machines (SVM), decision trees, and neural networks.

Regression methods let you build models from current data to predict future data. The models can then be updated as new data becomes available. If the data is only used once to create the model, then it is a batch method. A regression method that incorporates data as it becomes available is a recursive method.

Clustering finds natural groupings in data. Object recognition is an application of clustering methods. For example, if you want to find a car in an image, you look for data that is associated with the part of an image that is a car. While cars are of different shapes and sizes, they have many features in common. Clustering can also deal with different orientations and scalings.

The toolbox has many functions to support these areas and many that do not fit neatly into these categories. The Statistics and Machine Learning Toolbox provides professional tools that are seamlessly integrated into the MATLAB environment.

Computer Vision Toolbox

The MATLAB Computer Vision Toolbox provides functions for developing computer vision systems. The toolbox provides extensive support for video processing. It includes functions for feature detection and extraction. Prior to the extensive use of deep learning, feature detection was the approach for image identification. It also supports 3D vision and can process information from stereo cameras. 3D motion detection is supported.

Image Acquisition Toolbox

The MATLAB Image Acquisition Toolbox provides functions for connecting cameras directly into MATLAB without the need for intermediary software or using the apps that come with many cameras. You can use the package to interact with the sensors directly. Foreground and background acquisition is supported. The toolbox supports all major standards and hardware vendors. It makes it easier to design deep learning image processing software using real data. It allows control of cameras. The Image Acquisition Toolbox supports USB3 Vision, GigE Vision, and GenICam GenTL. You can connect to Velodyne LiDAR sensors, machine vision cameras, and frame grabbers, as well as high-end scientific and industrial devices. USB3 Vision gives you considerable control over the camera.

Parallel Computing Toolbox

The Parallel Computing Toolbox allows you to use multicore processors, Graphical Processing Units (GPUs), and computer clusters with your MATLAB software. It allows you to easily parallelize algorithms using high-level programming constructs like parallel `for` loops. Some functions in the Deep Learning Toolbox can take advantage of GPUs and parallel processing. There is an example of potential GPU use in Chapter 10. As almost every personal computer has a GPU, this can be a worthwhile addition to your MATLAB software.

Text Analytics Toolbox

The Text Analytics Toolbox provides algorithms and visualizations for working with text data. Models created with the toolbox can be used in applications such as sentiment analysis, predictive maintenance, and topic modeling. The toolbox includes tools for processing raw text from many sources. You can extract individual words, convert text into numerical representations, and build statistical models. This is a useful adjunct to deep learning.

2.2   MATLAB Open Source

MATLAB open source tools are a great resource for implementing machine learning. Machine learning and convex optimization packages are available. Universities are constantly producing new neural network toolsets. Much work is done in Python, but MATLAB is a very popular base for software development and AI work.

2.3 XOR Example

We'll give many examples of the Deep Learning Toolbox in subsequent chapters. We'll do one example just to get you going. This example doesn't even unlock a fraction of the power in the Deep Learning Toolbox. We will implement the XOR example which we also did in Chapter 1 using the Deep Learning Toolbox. The DLXOR.m script is shown in the following, using the MATLAB functions feedforwardnet, configure, train, and sim.

DLXOR.m

```
1   %% Use the Deep Learning Toolbox to create the XOR neural net
2
3   %% Create the network
4   % 2 layers
5   % 2 inputs
6   % 1 output
7
8   net = feedforwardnet(2);
9
10  % XOR Truth table
11  a     = [1 0 1 0];
12  b     = [1 0 0 1];
13  c     = [0 0 1 1];
14
15  % How many sets of inputs
16  n     = 600;
17
18  % This determines the number of inputs and outputs
19  x     = zeros(2,n);
20  y     = zeros(1,n);
21
22  % Create training pairs
23  for k = 1:n
24     j         = randi([1,4]);
25     x(:,k)    = [a(j); b(j)];
26     y(k)      = c(j);
27  end
28
29  net       = configure(net, x, y);
30  net.name  = 'XOR';
31  net       = train(net,x,y);
32  c         = sim(net,[a;b]);
33
34  fprintf('\n   a     b     c\n');
35  for k = 1:4
36     fprintf('%5.0f %5.0f %5.2f\n',a(k),b(k),c(k));
37  end
38
39  % This only works for feedforwardnet(2);
40  fprintf('\nHidden layer biases %6.3f %6.3f\n',net.b{1});
41  fprintf('Output layer bias    %6.3f\n',net.b{2});
```

```
42  fprintf('Input layer weights  %6.2f %6.2f\n',net.IW{1}(1,:));
43  fprintf('                      %6.2f %6.2f\n',net.IW{1}(2,:));
44  fprintf('Output layer weights %6.2f %6.2f\n',net.LW{2,1}(1,:));
45
46  fprintf('Hidden layer activation function %s\n',net.layers{1}.
        transferFcn);
47  fprintf('Output layer activation function %s\n',net.layers{2}.
        transferFcn);
```

Running the script produces the MATLAB GUI shown in Figure 2.1.

As you can see, we have two inputs, one hidden layer and one output layer. The diagram indicates that our hidden layer activation function is nonlinear, while the output layer is linear. The GUI is interactive, and you can study the learning process by clicking the buttons. For example, if you click the performance button, you get Figure 2.2. Just about everything in the network development is customizable. The GUI is a real-time display. You can watch the training in progress. If you just want to look at the layout, type `view(net)`.

The three major boxes in the GUI are Algorithms, Progress, and Plots. Under **Algorithms**, we have

- **Data Division** – Data Division divides the data into training, validation, and test sets. "Random" means that the division between the three categories is done randomly.
- **Training** – This shows the training method to be used.
- **Performance** – This says that the mean squared error (MSE) is used to determine how well the network works. Other methods, such as the maximum absolute error, could be used. Mean squared is useful because the error grows as the square of the deviation, meaning that large errors are more heavily weighted.
- **Calculations** – This shows that the calculations are done via a MEX file, that is, in a C or C++ program.

The **Progress** of the GUI is useful to watch during long training sessions. We are seeing it at the end.

- **Epoch** – This says five epochs were used. The range is 0 to 1000 epochs.
- **Time** – This gives you the clock time during training.
- **Performance** – This shows you the MSE performance during training.
- **Gradient** – This shows the gradient that shows the speed of training as discussed earlier.
- **Mu** – Mu is the control parameter for training the neural network.
- **Validation Checks** – This shows that no validation checks failed.

The last section is **Plots**. There are four figures we can study to understand the process.

Figure 2.2 shows the training performance as a function of epoch. The mean squared error is the criteria. The test, validation, and training sets have their own lines. In this training, all have the same values.

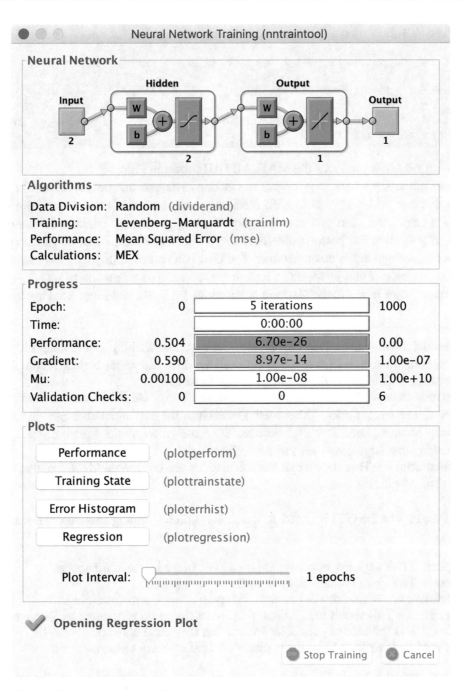

Figure 2.1: *Deep learning network GUI.*

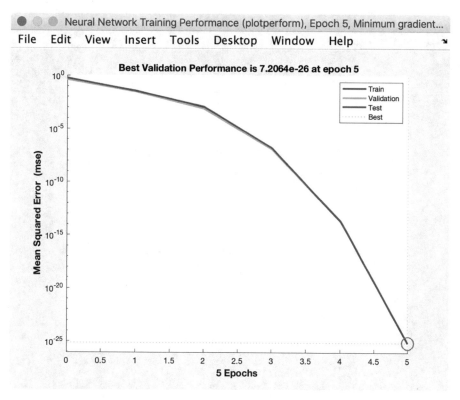

Figure 2.2: *Network training performance.*

Figure 2.3 shows the training state as a function of epoch. Five epochs are used. The titles show the final values in each plot. The top plot shows the progression of the gradient. It decreases with each epoch. The next shows mu decreasing linearly with epoch. The bottom plot shows that there were no validation failures during the training.

Figure 2.4 gives a training histogram. This shows the number of instances when one of the sets shows the error value on the x-axis. The bars are divided into training, validation, and test sets. Each number on the x-axis is a bin. Only three bins are occupied, in this case. The histogram shows that the training sets are more numerous than the validation or test sets.

Figure 2.5 gives a training regression. There are four subplots: one for training sets, one for validation sets, one for test sets, and one for all sets. There are only two targets, zero and one. The linear fit doesn't give much information in this case since we can only have a linear fit with two points. The plot title says we reached the minimum gradient after five epochs, that is, after passing all the cases through the training five times. The legend shows the data, the fit, and the Y=T plot which is the same as the linear in this system.

Typing

```
>> net = feedforwardnet(2);
```

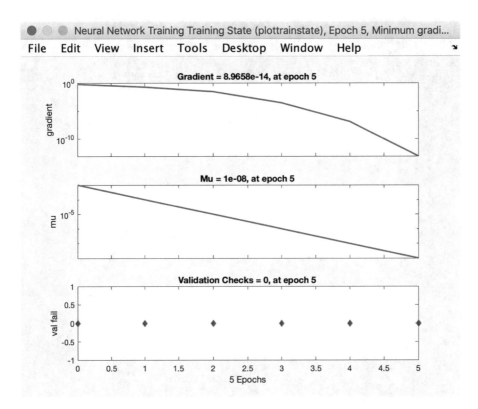

Figure 2.3: *Network training state.*

creates the neural network data structure which is quite flexible and complex. The "2" means two neurons in one layer. If we wanted two layers with two neurons each, we would type

```
>> net = feedforwardnet([2 2]);
```

We create 600 training sets. `net = configure(net, x, y);` configures the network. The configure function determines the number of inputs and outputs from the x and y arrays. The network is trained with `net = train(net,x,y);` and simulated with `c = sim(net,[a;b]);`. We extract the weights and biases from the cell arrays `net.IW`, `net.LW`, and `net.b`. "I" stands for input and "IW" for layer. Input is from the single input node to the two hidden nodes, and layer is from the two hidden nodes to the one output node.

Now the training sets are created randomly from the truth table. You can run this script many times, and usually you will get the right result, but not always. This is an example where it worked well:

```
>> DLXOR

    a     b    c
    1     1   0.00
```

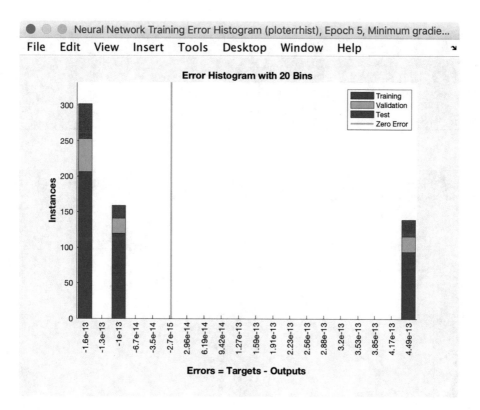

Figure 2.4: *Network training histogram.*

```
      0      0   0.00
      1      0   1.00
      0      1   1.00

Hidden layer biases   1.735 -1.906
Output layer bias     1.193
Input layer weights    -2.15    1.45
                       -1.83    1.04
Output layer weights  -1.16    1.30
Hidden layer activation function tansig
Hidden layer activation function purelin
```

`tansig` is a hyperbolic tangent sigmoid function.

Each run will result in different weights, even when the network gives the correct results. For example:

```
>> DLXOR

      a      b   c
      1      1   0.00
      0      0  -0.00
      1      0   1.00
```

```
    0      1  1.00

Hidden layer biases   4.178   0.075
Output layer bias     -1.087
Input layer weights   -4.49   -1.36
                      -3.79   -3.67
Output layer weights   2.55   -2.46
```

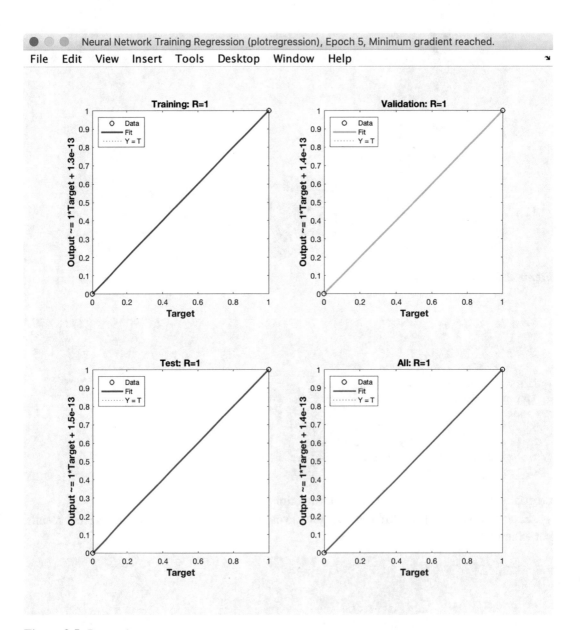

Figure 2.5: *Regression.*

There are many possible sets of weights and biases because the weight/bias sets are not unique. Note that the 0.00 are really not 0. This means that used operationally, we would need to set a threshold such as

```
1  if( abs(c) < tol )
2     c = 0;
3  end
```

You might be interested in what happens if we add another layer to the network, by creating it with `net = feedforwardnet([2 2])`. Figure 2.6 shows the network in the GUI.

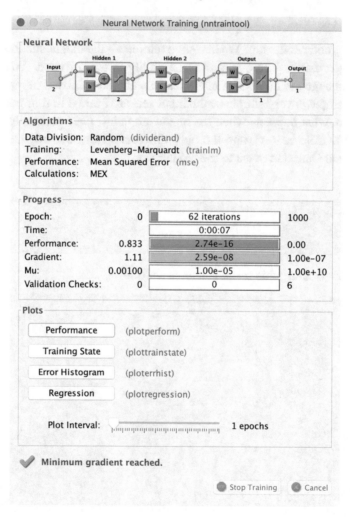

Figure 2.6: *Deep learning network GUI with two hidden layers.*

The additional hidden layer makes it easier for the neural net to fit the data from which it is learning. On the left are the two inputs, a and b. In each hidden layer, there is a weight w and bias b. Weights are always needed, but biases are not always used. Both hidden layers have nonlinear activation functions. The output layer produces the one output using a linear activation function.

```
>> DLXOR
     a     b     c
     1     1   0.00
     0     0   0.00
     1     0   1.00
     0     1   1.00
```

This produces good results too. We haven't explored all the diagnostic tools available when using `feedforwardnet`. There is a lot of flexibility in the software. You can change activation functions, change the number of hidden layers, and customize it in many different ways. This particular example is very simple as the input sets are limited to four possibilities.

We can explore what happens when the inputs are noisy, not necessarily all ones or zeros. We do this in DLXORNoisy.m, and the only difference from the original script is in lines 33–35 where we add Gaussian noise to the inputs:

DLXORNoisy.m

```
33  a          = a + 0.01*randn(1,4);
34  b          = b + 0.01*randn(1,4);
35  c          = sim(net,[a;b]);
```

The output from running this script is shown as follows:

```
>> DLXORNoisy

     a      b      c
 0.991  1.019 -0.003
 0.001 -0.005 -0.002
 0.996  0.009  0.999
-0.001  1.000  1.000

Hidden layer biases -1.793  2.135
Output layer bias    -1.158
Input layer weights      1.70    1.54
                         1.80    1.52
Output layer weights -1.11    1.15
Hidden layer activation function tansig
Output layer activation function purelin
```

As one might expect, the outputs are not exactly one or zero.

2.4   Training

The neural net is a nonlinear system due to the nonlinear activation functions. The Levenberg-Marquardt training algorithm is one way of solving a nonlinear least squares problem. This algorithm only finds a local minimum which may or may not be a global minimum. Other algorithms, such as Genetic Algorithms, Downhill Simplex, Simulated Annealing, and so on, could also be used for finding the weights and biases. To achieve second-order training speeds, one has to compute the Hessian matrix. The Hessian matrix is a square matrix of second-order partial derivative of a scalar-valued function. Suppose we have a nonlinear function:

$$f(x_1, x_2) \tag{2.1}$$

then the Hessian is

$$H = \begin{bmatrix} \frac{\partial^2 f}{\partial x_1^2} & \frac{\partial^2 f}{\partial x_1 \partial x_2} \\ \frac{\partial^2 f}{\partial x_2 \partial x_1} & \frac{\partial^2 f}{\partial x_2^2} \end{bmatrix} \tag{2.2}$$

x_k are weights and biases. This can be very expensive to compute. In the Levenberg-Marquardt algorithm, we make an approximation:

$$H = J^T J \tag{2.3}$$

where

$$J = \begin{bmatrix} \frac{\partial f}{\partial x_1} & \frac{\partial f}{\partial x_2} \end{bmatrix} \tag{2.4}$$

The approximate Hessian is

$$H = \begin{bmatrix} (\frac{\partial f}{\partial x_1})^2 & \frac{\partial f}{\partial x_1} \frac{\partial f}{\partial x_2} \\ \frac{\partial f}{\partial x_1} \frac{\partial f}{\partial x_2} & (\frac{\partial f}{\partial x_2})^2 \end{bmatrix} \tag{2.5}$$

This is an approximation of the second derivative that only requires direct computation of first derivatives. This approach reduces the overall computational burden by eliminating the need to compute the second derivatives. The gradient is

$$g = J^T e \tag{2.6}$$

where e is a vector of errors. The Levenberg-Marquardt uses the following algorithm to update the weights and biases:

$$x_{k+1} = x_k - \left[J^T J + \mu I \right]^{-1} J^T e \tag{2.7}$$

I is the identity matrix (a matrix with all diagonal elements equal to 1). If the parameter μ is zero, this is Newton's method. With a large μ, this becomes gradient descent which is faster. Thus, μ is a control parameter. After a successful step, we decrease μ since we are in less need of the advantages of the faster gradient descent.

Why are gradients so important and why can they get us into trouble? Figure 2.7 shows a curve with a local and global minimum. If our search first enters the local minimum, the gradient is steep and will drive us to the bottom from which we might not get out. Thus, we would not have found the best solution.

The cost can be very complex even for simple problems. A famous problem that can give insight into the problem is Zermelo's problem which is discussed in Section 2.5.

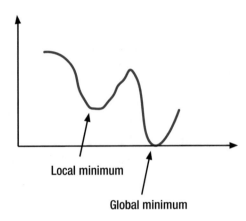

Figure 2.7: *Local and global minimums.*

2.5 Zermelo's Problem

Insight into issues of global optimization can be obtained by examining Zermelo's problem [6]. Zermelo's problem is a 2D trajectory problem of a vehicle at constant speed in a velocity field in which the velocity is a function of position, for example, a ship with a given maximum speed moving through strong currents. The magnitude and direction of the currents in each axis (u,v) are functions of the position: $u(x,y)$ and $v(x,y)$. The goal is to steer the ship to find a minimum time trajectory between two points. An analytical solution is possible for the case where only the current in the x direction is nonzero, and it is a linear function of the y position of the vehicle. The equations of motion f are

$$
\begin{aligned}
\dot{x} &= V\cos\theta + u(x,y) = V\cos\theta - V\frac{y}{h} \\
\dot{y} &= V\sin\theta + v(x,y) = V\sin\theta
\end{aligned}
\tag{2.8}
$$

V is the velocity of the vehicle relative to the current which is constant, and θ is the angle of the vehicle relative to the x-axis and is the control in the problem. The problem has a characteristic dimension of h. The Hamiltonian of the system is

$$
H = \lambda_x(V\cos\theta + u) + \lambda_y(V\sin\theta + v) + 1
\tag{2.9}
$$

The Hamiltonian is the cost that we want to minimize. The solution of the optimal control problem requires the solution of the following equations:

$$
\dot{x} = f(x,u,t)
\tag{2.10}
$$

$$
\dot{\lambda}(t)^T = -\frac{\partial H}{\partial x}
\tag{2.11}
$$

$$
\frac{\partial H}{\partial u} = 0
\tag{2.12}
$$

If the final time is not constrained, and the Hamiltonian is not an explicit function of time, then

$$
H(t) = 0
\tag{2.13}
$$

The costate equations are

$$\dot{\lambda} = -\frac{\partial f}{\partial x}^T \lambda \qquad (2.14)$$

where the boundary conditions of the partials with respect to the vector x are unknown. A costate is a Lagrange multiplier associated with the state equations. The states are a constraint on the problem, that is, the system states are forced to evolve in a way specified by the differential equations for the system, that is, the state equations. The costates represent the marginal cost of violating those constraints. The optimality condition from (2.12) becomes

$$0 = \frac{\partial f}{\partial u}^T \lambda \qquad (2.15)$$

where the subscript denotes the partial with respect to the control vector u which provides a relationship between the controls and the costates.

Applying the costate equations (2.14), we first compute the partials of the state equations:

$$\frac{\partial f}{\partial x} = \begin{bmatrix} 0 & -V/h \\ 0 & 0 \end{bmatrix} \qquad (2.16)$$

so applying the transpose to the partials matrix and substituting in, we then have the costate derivatives:

$$\dot{\lambda} = -\begin{bmatrix} 0 & 0 \\ -V/h & 0 \end{bmatrix} \begin{bmatrix} \lambda_x \\ \lambda_y \end{bmatrix} = \begin{bmatrix} 0 \\ \lambda_x \frac{V}{h} \end{bmatrix} \qquad (2.17)$$

We then compute the partials of the state equations with respect to the control:

$$\frac{\partial f}{\partial u} = \begin{bmatrix} -V \sin \theta \\ V \cos \theta \end{bmatrix} \qquad (2.18)$$

so that the control angle θ can then be computed from the optimality condition in (2.15):

$$\begin{bmatrix} -V \sin \theta & V \cos \theta \end{bmatrix} \begin{bmatrix} \lambda_x \\ \lambda_y \end{bmatrix} = 0 \qquad (2.19)$$

$$\tan \theta = \frac{\lambda_y}{\lambda_x}$$

The preceding equations are used for the indirect optimization method. We can also compute the analytical solution for this problem when the final position is the origin (0,0). The optimal control angle as a function of the current position is expressed as

$$\frac{y}{h} = \sec \theta - \sec \theta_f \qquad (2.20)$$

$$\frac{x}{h} = -\frac{1}{2} \left[\sec \theta_f (\tan \theta_f - \tan \theta) - \tan \theta_f (\sec \theta_f - \sec \theta) + \log \frac{\tan \theta_f + \sec \theta_f}{\tan \theta + \sec \theta} \right] \qquad (2.21)$$

where θ_f is the final control angle and log is $\log_e$. These equations enable us to solve for the initial and final control angles θ_0 and θ_f given an initial position.

The cost, while appearing simple, produces a very complex surface as shown in Figure 2.8. There are very flat regions and then a series of deep valleys.

For each method being tested, the optimization parameters have been chosen, by a certain amount of trial and error, to get the best results. The final λ vector is different in each case. However, the control is determined by the ratio, so the magnitudes are not important. Table 2.1 gives the analytical and numerical solutions for the problem. The initial conditions are $[3.66; -1.86]$, and the final conditions are $[0; 0]$.

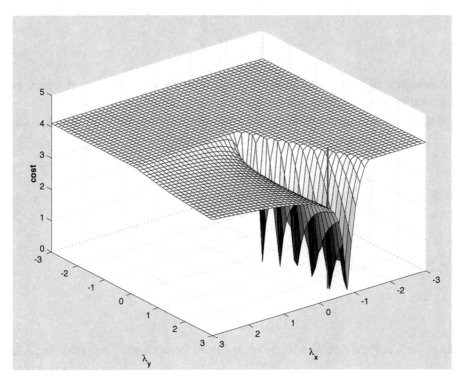

Figure 2.8: *Zermelo's problem cost.*

Table 2.1: *Solutions.*

Test	λ_x	λ_y	Ratio
Analytical	-0.5	1.866025	-0.26795
Downhill Simplex	-0.65946	2.9404	-0.22428
Simulated Annealing	-0.68652	2.4593	-0.27915
Genetic Algorithm	-0.78899	2.9404	-0.26833

CHAPTER 3

■ ■ ■

Finding Circles with Deep Learning

3.1 Introduction

Finding circles is a classification problem. Given a set of geometric shapes, we want the deep learning system to classify a shape as either a circle or something else. This is much simpler than classifying faces or digits. It is a good way to determine how well your classification system works. We will apply a convolutional network to the problem as it is the most appropriate for classifying image data.

In this chapter, we will first generate a set of image data. This will be a set of ellipses, a subset of which will be circles. Then we will build the neural net, using convolution, and train it to identify the circles. Finally, we will test the neural network and try some different options for training options and layer architecture.

3.2 Structure

The convolutional network consists of multiple layers. Each layer has a specific purpose. The layers may be repeated with different parameters as part of the convolutional network. The layer types we will use are

1. imageInputLayer

2. convolution2dLayer

3. batchNormalizationLayer

4. reluLayer

5. maxPooling2dLayer

6. fullyConnectedLayer

© Michael Paluszek, Stephanie Thomas, Eric Ham 2022
M. Paluszek et al., *Practical MATLAB Deep Learning*,
https://doi.org/10.1007/978-1-4842-7912-0_3

7. softmaxLayer

8. classificationLayer

You can have multiple layers of each type of layer. Some convolutional nets have hundreds of layers. Krizhevsky [22] and Bai [2] give guidelines for organizing the layers. Studying the loss in the training and validation can guide you to improving your neural network.

3.2.1 imageInputLayer

This tells the network the size of the images. For example:

```
1  layer = imageInputLayer([28 28 3]);
```

says the image is RGB and 28 by 28 pixels.

3.2.2 convolution2dLayer

Convolution is the process of highlighting expected features in an image. This layer applies sliding convolutional filters on an image to extract features. You can specify the filters and the stride. Convolution is a matrix multiplication operation. You define the size of the matrices and their contents. For most images, like images of faces, you need multiple filters. Some types of filters are

1. **Blurring filter** – ones(3,3)/9

2. **Sharpening filter** – [0 -1 0;-1 5 -1;0 -1 0]

3. **Horizontal Sobel filter for edge detection** – [-1 -2 -1; 0 0 0; 1 2 1]

4. **Vertical Sobel filter for edge detection** – [-1 0 1;-2 0 2;-1 0 1]

We create an n by n mask that we apply to an m by m matrix of data where m is greater than n. We start in the upper-left corner of the matrix, as shown in Figure 3.1. We multiply the mask times the corresponding elements in the input matrix and sum all the elements using two calls to sum. That is the first element of the convolved output. We then move it column by column until the highest column of the mask is aligned with the highest column of the input matrix. We then return it to the first column and increment the row. We continue until we have traversed the entire input matrix and our mask is aligned with the maximum row and maximum column.

The mask represents a feature. In effect, we are seeing if the feature appears in different areas of the image. Here is an example: we have a 2 by 2 mask with an L. The script Convolution.m demonstrates convolution.

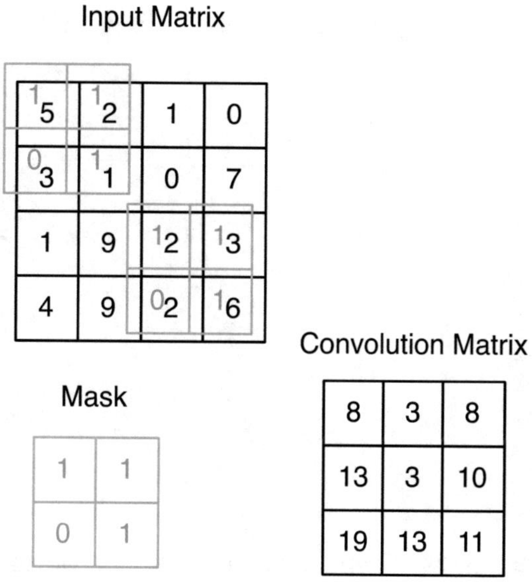

Input Matrix

Mask

Convolution Matrix

Figure 3.1: *Convolution process showing the mask at the beginning and end of the process.*

Convolution.m

```
1   %% Demonstrate convolution
2
3   filter = [1 0;1 1]
4   image  = [0 0 0 0 0 0;...
5            0 0 0 0 0 0;...
6            0 0 1 0 0 0;...
7            0 0 1 1 0 0;...
8            0 0 0 0 0 0]
9
10  out = zeros(3,3);
11
12  for k = 1:4
13    for j = 1:4
14      g = k:k+1;
15      f = j:j+1;
16      out(k,j) = sum(sum(filter.*image(g,f)));
17    end
18  end
19
20  out
```

The number 3 appears in the input matrix where the "L" is in the image.

```
>> Convolution
filter =
       1       0
       1       1
image =
       0       0       0       0       0       0
       0       0       0       0       0       0
       0       0       1       0       0       0
       0       0       1       1       0       0
       0       0       0       0       0       0
out  =
       0       0       0       0
       0       1       1       0
       0       1       3       1
       0       0       1       1
```

We can have multiple masks. There is one bias and one weight for each element of the mask for each feature. In this case, the convolution works on the image itself. Convolutions can also be applied to the output of other convolutional layers or pooling layers. Pooling layers further condense the data. In deep learning, the masks are determined as part of the learning process. Each pixel in a mask has a weight and may have a bias; these are computed from the learning data. Convolution should be highlighting important features in the data. Subsequent convolution layers narrow down features. The MATLAB function has two inputs: the filterSize, specifying the height and width of the filters as either a scalar or an array of [h w], and numFilters, the number of filters.

3.2.3 batchNormalizationLayer

A batch normalization layer normalizes each input channel across a mini-batch. It automatically divides up the input channel into mini-batches, which are subsets of the entire batch. This reduces the sensitivity to the initialization.

3.2.4 reluLayer

reluLayer is a layer that uses the rectified linear unit (ReLU) activation function:

$$f(x) = \begin{cases} x & x >= 0 \\ 0 & x < 0 \end{cases} \tag{3.1}$$

Its derivative is

$$\frac{df}{dx} = \begin{cases} 1 & x >= 0 \\ 0 & x < 0 \end{cases} \tag{3.2}$$

This is very fast to compute. It says that the neuron is only activated for positive values, and the activation is linear for any value greater than zero. You can adjust the activation point with a bias. This code snippet generates a plot of reluLayer:

```
x = linspace(-8,8);
y = x;
y(y<0) = 0;
PlotSet(x,y,'x label','Input','y label','reluLayer','plot title','
    reluLayer')
```

Figure 3.2 shows the activation function. An alternative is a leaky reluLayer where the value is not zero for negative input values.

Now the difference is the y computation in the following snippet:

```
x = linspace(-8,8);
y = x;
y(y<0) = 0.01*x(y<0);
PlotSet(x,y,'x label','Input','y label','reluLayer','plot title','leaky
    reluLayer')
```

Figure 3.3 shows the leaky function. Below zero, it has a slight slope.

A leaky ReLU layer solves the dead ReLU problem where the network stops learning because the inputs to the activation problem are below zero, or whatever the threshold might be. It should let you worry a bit less on how you initialize the network.

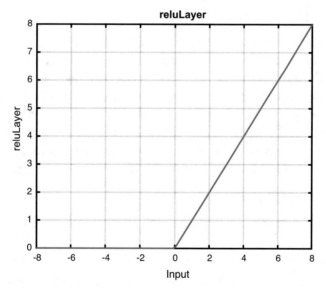

Figure 3.2: *reluLayer.*

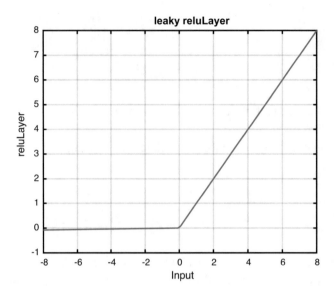

Figure 3.3: *Leaky reluLayer.*

3.2.5 maxPooling2dLayer

`maxPooling2dLayer` creates a layer that breaks the 2D input into rectangular pooling regions, and outputs the maximum value of each region. The input `poolSize` specifies the width and height of a pooling region. `poolSize` can have one element (for square regions) or two for rectangular regions. This is a way to reduce the number of inputs that need to be evaluated. Typical images have to be a mega-pixel, and it is not practical to use all pixels as inputs. Furthermore, most images, or two-dimensional entities of any sort, don't really have enough information to require finely divided regions. You can experiment with pooling and see how it works for your application. An alternative is `averagePooling2dLayer`.

3.2.6 fullyConnectedLayer

The fully connected layer connects all of the inputs to the outputs with weights and biases. For example:

```
1  layer = fullyConnectedLayer(10);
```

creates ten outputs from any number of inputs. You don't have to specify the inputs. Effectively, this is the equation:

$$y = ax + b \tag{3.3}$$

If there are m inputs and n outputs, b is a column bias matrix of length n and a is n by m.

3.2.7 softmaxLayer

Softmax finds a maximum of a set of values using the logistic function. The softmax is the maximum value of the set

$$p_k = \frac{e_k^q}{\sum e^{q_k}} \tag{3.4}$$

```
>> q = [1,2,3,4,1,2,3]

q =

    1    2    3    4    1    2    3

>> d = sum(exp(q));
>> p = exp(q)/d

p =

    0.0236    0.0643    0.1747    0.4748    0.0236    0.0643    0.1747
```

In this case, the maximum is element four in both cases. This is just a method of smoothing the inputs. Softmax is used for multiclass classification because it guarantees a well-behaved probability distribution. Well behaved means that the sum of the probabilities is one.

3.2.8 classificationLayer

A classification layer computes the cross-entropy loss for multiclass classification problems with mutually exclusive classes. Let us define loss. The loss is the sum of the errors in training the neural net. It is not a percentage. For classification, the loss is usually the negative log likelihood, which is

$$L(y) = -\log(y) \tag{3.5}$$

where y is the output of the softmax layer.

For regression, it is the residual sum of squares. A high loss means a bad fit.

Cross-entropy loss is a useful metric when an item being classified can only be in one class. The number of classes is inferred from the output of the previous layer. In this problem, we have only two classes, circle and ellipse, so the number of outputs of the previous layer must be two. Cross-entropy is the distance between the original probability distribution and what the model believes it should be. It is defined as

$$H(y,p) = -\sum_i y_i \log p_i \tag{3.6}$$

where i is the index for the class. It is a widely used replacement for mean squared error. It is used in neural nets where softmax activations are in the output layer.

3.2.9 Structuring the Layers

For our first net to identify circles, we will use the following set of layers. The first layer is the input layer, for the 32×32 images. These are relatively low-resolution images. You can visually determine which are ellipses or circles, so we would expect the neural network to be able to do the same. Nonetheless, the size of the input images is an important consideration. In our case, our images are tightly cropped around the shape. In a more general problem, the subject of interest, a cat, for example, might be in a general setting.

We use a `convolution2dLayer`, `batchNormalizationLayer`, and `reluLayer` in sequence, with a pool layer in between. There are three sets of convolution layers, each with an increasing number of filters. The output set of layers consists of a `fullyConnectedLayer`, `softmaxLayer`, and, finally, the `classificationLayer`.

EllipsesNeuralNet.m

```
34  %% Define the layers for the net
35  % This gives the structure of the convolutional neural net
36  layers = [
37      imageInputLayer(size(img))
38
39      convolution2dLayer(3,8,'Padding','same')
40      batchNormalizationLayer
41      reluLayer
42
43      maxPooling2dLayer(2,'Stride',2)
44
45      convolution2dLayer(3,16,'Padding','same')
46      batchNormalizationLayer
47      reluLayer
48
49      maxPooling2dLayer(2,'Stride',2)
50
51      convolution2dLayer(3,32,'Padding','same')
52      batchNormalizationLayer
53      reluLayer
54
55      fullyConnectedLayer(2)
56      softmaxLayer
57      classificationLayer
58      ];
```

3.3   Generating Data: Ellipses and Circles

3.3.1 Problem

We want to generate images of ellipses and circles of arbitrary sizes and with different thicknesses in MATLAB.

3.3.2 Solution

Write a MATLAB function to draw circles and ellipses and extract image data from the figure window. Our function will create a set of ellipses and a fixed number of perfect circles as specified by the user. The actual plot and the resulting downsized image will both be shown in a figure window so you can track progress and verify that the images look as expected.

3.3.3 How It Works

This is implemented in `GenerateEllipses.m`. The output of the function is a cell array with both the ellipse data and a set of image data obtained from a MATLAB figure using `getframe`. The function also outputs the type of image, that is, the "truth" data.

GenerateEllipses.m

```
1   %% GENERATEELLIPSES Generate random ellipses
2   %% Form
3   %   [d, v] = GenerateEllipses(a,b,phi,t,n,nC,nP)
4   %% Description
5   % Generates random ellipses given a range of sizes and max rotation.
       The number
6   % of ellipses and circles must be specified; the total number generated
       is their
7   % sum. Opens a figure which displays the ellipse images in an animation
       after
8   % they are generated.
9   %% Inputs
10  %   a    (1,2) Range of a sizes of ellipse
11  %   b    (1,2) Range of b sizes of ellipse
12  %   phi  (1,1) Max rotation angle of ellipse
13  %   t    (1,1) Max line thickness in the plot of the circle
14  %   n    (1,1) Number of ellipses
15  %   nC   (1,1) Number of circles
16  %   nP   (1,1) Number of  pixels, image is nP by nP
17  %
18  %% Outputs
19  %   d        {:,2} Ellipse data and image frames
20  %   v        (1,:) Boolean for circles, 1 (circle) or 0 (ellipse)
```

The first section of the code generates random ellipses and circles. They are all centered in the image.

GenerateEllipses.m

```
31  nE      = n+nC;
32  d       = cell(nE,2);
33  r       = 0.5*(mean(a) + mean(b))*rand(1,nC)+a(1);
34  a       = (a(2)-a(1))*rand(1,n) + a(1);
35  b       = (b(2)-b(1))*rand(1,n) + b(1);
36  phi     = phi*rand(1,n);
37  cP      = cos(phi);
38  sP      = sin(phi);
39  theta   = linspace(0,2*pi);
40  c       = cos(theta);
41  s       = sin(theta);
42  m       = length(c);
43  t       = 0.5+(t-0.5)*rand(1,nE);
44  aMax    = max([a(:);b(:);r(:)]);
45
46  % Generate circles
47  for k = 1:nC
48    d{k,1} = r(k)*[c;s];
49  end
50
51  % Generate ellipses
52  for k = 1:n
53    d{k+nC,1} = [cP(k) sP(k);-sP(k) cP(k)]*[a(k)*c;b(k)*s];
54  end
55
56  % True if the object is a circle
57  v       = zeros(1,nE);
58  v(1:nC) = 1;
```

The next section produces a 3D plot showing all the ellipses and circles. This is just to show you what you have produced. The code puts all the ellipses between $z \pm 1$. You might want to adjust this when generating larger numbers of ellipses.

```
60  % 3D Plot
61  NewFigure('Ellipses');
62  z   = -1;
63  dZ = 2*abs(z)/nE;
64  o   = ones(1,m);
65  for k = 1:length(d)
66    z   = z + dZ;
67    zA  = z*o;
68    plot3(d{k}(1,:),d{k}(2,:),zA,'linewidth',t(k));
69    hold on
70  end
71  grid on
72  rotate3d on
```

The next section converts the frames to nP by nP sized images in grayscale. We set the figure and the axis to be square, and set the axis to 'equal', so that the circles will have the correct aspect ratio and in fact be circular in the images. Otherwise, they would also appear as ellipses, and our neural net would not be able to categorize them. This code block also draws the resulting resized image on the right-hand side of the window, with a title showing the current step. There is a brief pause between each step. In effect, it is an animation that serves to inform you of the script's progress.

```
74  % Create images - this might take a while for a lot of images
75  f = figure('Name','Images','visible','on','color',[1 1 1]);
76  ax1 = subplot(1,2,1,'Parent', f, 'box', 'off','color',[1 1 1] );
77  ax2 = subplot(1,2,2,'Parent',f); grid on;
78  for k = 1:length(d)
79    % Plot the ellipse and get the image from the frame
80    plot(ax1,d{k}(1,:),d{k}(2,:),'linewidth',t(k),'color','k');
81    axis(ax1,'off'); axis(ax1,'equal');
82    axis(ax1,aMax*[-1 1 -1 1])
83    frame  = getframe(ax1); % this call is what takes time
84    imSmall = rgb2gray(imresize(frame2im(frame),[nP nP]));
85    d{k,2} = imSmall;
86    % plot the resulting scaled image in the second axes
87        imagesc(ax2,d{k,2});
88    axis(ax2,'equal')
89        colormap(ax2,'gray');
90    title(ax2,['Image ' num2str(k)])
91        set(ax2,'xtick',1:nP)
92        set(ax2,'ytick',1:nP)
93        colorbar(ax2)
94        pause(0.2)
95  end
96  close(f)
```

The conversion is rgb2gray(imresize(frame2im(frame), [nP nP])), which performs these steps:

1. Get the frame with frame2im

2. Resize to nP by nP using imresize

3. Convert to grayscale using rgb2gray

Note that the image data originally ranges from 0 (black) to 255 (white), but is averaged to lighter gray pixels during the resize operation. The colorbar in the progress window shows you the span of the output image. The image looks black as before since it is plotted with imagesc, which automatically scales the image to use the entire colormap – in this case, the gray colormap.

The built-in demo generates ten ellipses and five circles.

```
100  function Demo
101
102  a   = [0.5 1];
103  b   = [1 2];
104  phi = pi/4;
105  t   = 3;
106  n   = 10;
107  nC  = 5;
108  nP  = 32;
109
110  GenerateEllipses(a,b,phi,t,n,nC,nP);
```

Figure 3.4 shows the generated ellipses and the first image displayed.

The script `CreateEllipses.m` generates 100 ellipses and 100 circles and stores them in the `Ellipses` folder along with the type of each image. Note that we have to do a small trick with the filename. If we simply append the image number to the filename, 1, 2, 3, ... 200, the images will not be in this order in the datastore; in alphabetical order, the images would be sorted as 1, 10, 100, 101, and so on. In order to have the filenames in alphabetical order match the order we are storing with the type, we generate a number a factor of 10 higher than the number of images and add it to the image index before appending it to the file. Now we have image 1001, 1002, and so on.

CreateEllipses.m

```
1  %% Create ellipses to train and test the deep learning algorithm
2  % The ellipse images are saved  as jpegs in the folder Ellipses.
3
4  % Parameters
5  nEllipses = 1000;
```

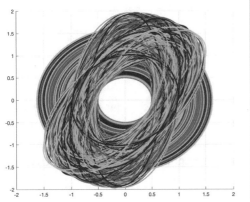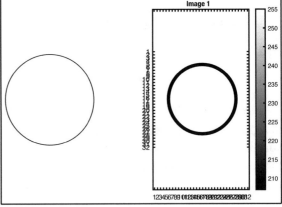

Figure 3.4: *Ellipses and a resulting image.*

```
6   nCircles   = 1000;
7   nBits      = 32;
8   maxAngle   = pi/4;
9   rangeA     = [0.5 1];
10  rangeB     = [1 2];
11  maxThick   = 3.0;
12  tic
13  [s, t] = GenerateEllipses(rangeA,rangeB,maxAngle,maxThick,nEllipses,
            nCircles,nBits);
14  toc
15  cd Ellipses
16  kAdd = 10^ceil(log10(nEllipses+nCircles)); % to make a serial number
17  for k = 1:length(s)
18    imwrite(s{k,2},sprintf('Ellipse%d.jpg',k+kAdd));
19  end
20
21  % Save the labels
22  save('Type','t');
23  cd ..
```

The graphical output is shown in Figure 3.5. It first displays the 100 circles and then the 100 ellipses. It takes some time for the script to generate all the images.

If you open the resulting jpegs, you will see that they are in fact 32×32 images with gray circles and ellipses.

This recipe provides the data that will be used for the deep learning examples in the following sections. You must run CreateEllipses.m before you can run the neural net examples.

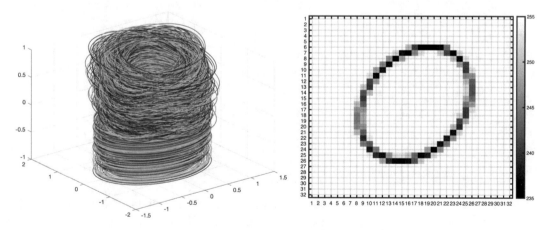

Figure 3.5: *Ellipses and a resulting image. 100 circles and 100 ellipse images are stored.*

3.4   Training and Testing

3.4.1 Problem

We want to train and test our deep learning algorithm on a wide range of ellipses and circles.

3.4.2 Solution

The script that creates, trains, and tests the net is `EllipsesNeuralNet.m`.

3.4.3 How It Works

First, we need to load in the generated images. The script in Recipe 3.3 generates 200 files. Half are circles and half ellipses. We will load them into an image datastore. We display a few images from the set to make sure we have the correct data and it is tagged correctly – that is, that the files have been correctly matched to their type, circle (1) or ellipse (0).

EllipsesNeuralNet.m

```
7   %% Get the images
8   cd Ellipses
9   type = load('Type');
10  cd ..
11  t    = categorical(type.t);
12  imds = imageDatastore('Ellipses','labels',t);
13
14  labelCount = countEachLabel(imds);
15
16  % Display a few ellipses
17  NewFigure('Ellipses')
18  n = 4;
19  m = 5;
20  ks = sort(randi(length(type.t),1,n*m)); % random selection
21  for i = 1:n*m
22          subplot(n,m,i);
23          imshow(imds.Files{ks(i)});
24      title(sprintf('Image %d: %d',ks(i),type.t(ks(i))))
25  end
26
27  % We need the size of the images for the input layer
28  img = readimage(imds,1);
```

Once we have the data, we need to create the training and testing sets. We have 100 files with each label (0 or 1, for an ellipse or circle). We create a training set of 80% of the files and reserve the remaining as a test set using `splitEachLabel`. Labels could be names, like "circle" and "ellipse." You are generally better off with descriptive "labels." After all, a 0 or 1 could mean anything. The MATLAB software handles many types of labels.

EllipsesNeuralNet.m

```
30  % Split the data into training and testing sets
31  fracTrain            = 0.8;
32  [imdsTrain,imdsTest] = splitEachLabel(imds,fracTrain,'randomize');
```

The layers of the net are defined as in the previous recipe. The next step is training. The `trainNetwork` function takes the data, set of layers, and options, runs the specified training algorithm, and returns the trained network. This network is then invoked with the `classify` function, as shown later in this recipe. This network is a series network. The network has other methods which you can read about in the MATLAB documentation.

EllipsesNeuralNet.m

```
60  %% Training
61  % The mini-batch size should be less than the data set size; the mini-
       batch is
62  % used at each training iteration to evaluate gradients and update the
       weights.
63  options = trainingOptions('sgdm', ...
64      'InitialLearnRate',0.01, ...
65      'MiniBatchSize',16, ...
66      'MaxEpochs',5, ...
67      'Shuffle','every-epoch', ...
68      'ValidationData',imdsTest, ...
69      'ValidationFrequency',2, ...
70      'Verbose',false, ...
71      'Plots','training-progress');
```

Figure 3.6 shows some of the ellipses used in the testing and training. They were obtained randomly from the set using `randi`.

The training options need explanation. This is a subset of the parameter pairs available for `trainingOptions`. The first input to the function, `'sgdm'`, specifies the training method. There are three to choose from:

1. `'sgdm'`   – Stochastic gradient descent with momentum

2. `'adam'`   – Adaptive moment estimation (ADAM)

3. `'rmsprop'` – Root mean square propagation (RMSProp)

The `'InitialLearnRate'` is the initial speed of learning. Higher learn rates mean faster learning, but the training may get stuck in a suboptimal point. The default rate for `sgdm` is 0.01. `'MaxEpochs'` is the maximum number of epochs to be used in the training. In each epoch, the training sees the entire training set, in batches of `MiniBatchSize`. The number of iterations in each epoch is therefore determined by the amount of data in the set and the `MiniBatchSize`. We are using a smaller dataset, so we reduce the `MiniBatchSize` from the default of 128 to 16, which will give us ten iterations per epoch. `'Shuffle'` tells the training how often to shuffle the training data. If you don't shuffle, the data will always be

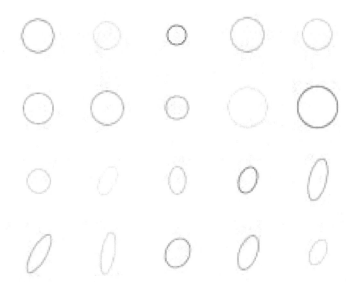

Figure 3.6: *A subset of the ellipses used in the training and testing.*

used in the same order. Shuffling should improve the accuracy of the trained neural network. `'ValidationFrequency'` is how often, in number of iterations, `'ValidationData'` is used to verify the trained neural network. This validation data will be the data we reserved for testing when using `splitEachLabel`. The default frequency is every 30 iterations. We can use a validation frequency for our small problem of one, two, or five iterations. `'Verbose'` means print out status information to the command window. `'Plots'` only has the option `'training-progress'` (besides `'none'`). This is the plot you see in this chapter.

The training window runs in real time with the training process. The window is shown in Figure 3.7. Our network starts with a 50% training accuracy (blue line) since we only have two classes, circles and ellipses. Our accuracy approaches 100% in just five epochs, indicating that our classes of images are readily distinguishable.

The loss plot shows how well we are doing. The lower the loss, the better the neural net. The loss plot approaches zero as the accuracy approaches 100%. In this case, the validation data loss (black dashed line) and the training data loss (red line) are about the same. This indicates good fitting of the neural net with the data. If the validation loss is greater than the training loss, the neural net is overfitting the data. Overfitting happens when you have an overly complex neural network. You can fit the training data, but it may not perform very well with new data, such as the validation data. For example, if you have a system which really is linear, and you fit it to a cubic equation, it might fit the data well but doesn't model the real system. If the training loss is greater than the validation data loss, your neural net is underfitting. Underfitting happens when your neural net is too simple. The goal is to make both zero.

Finally, we test the net. Remember that this is a classification problem. An image is either an ellipse or a circle. We therefore use `classify` to implement the network. `predLabels` is the output of the net, that is, the predicted labels for the test data. This is compared to the truth labels from the datastore to compute an accuracy.

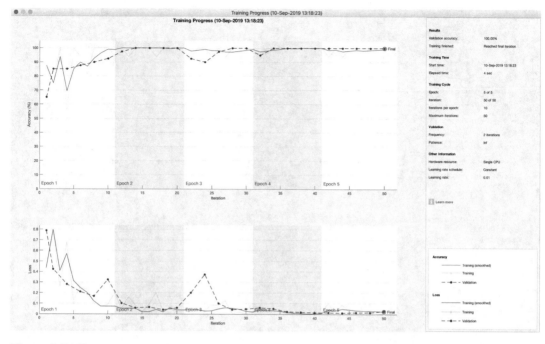

Figure 3.7: *The training window with a learn rate of 0.01. The top plot is the accuracy expressed as a percentage. The bottom plot shows the loss. As the accuracy increases, the loss decreases. The plots show the results with the training set and the validation set. The elapsed time is three seconds.*

EllipsesNeuralNet.m

```
75
76   %% Test the neural net
77   predLabels  = classify(net,imdsTest);
78   testLabels  = imdsTest.Labels;
79
80   accuracy = sum(predLabels == testLabels)/numel(testLabels);
```

The output of the testing is shown in the following. The accuracy of this run was 97.50%. On some runs, the net reaches 100%.

```
>> EllipsesNeuralNet

ans =

  Figure (1: Ellipses) with properties:

      Number: 1
        Name: 'Ellipses'
       Color: [0.9400 0.9400 0.9400]
    Position: [560 528 560 420]
       Units: 'pixels'
```

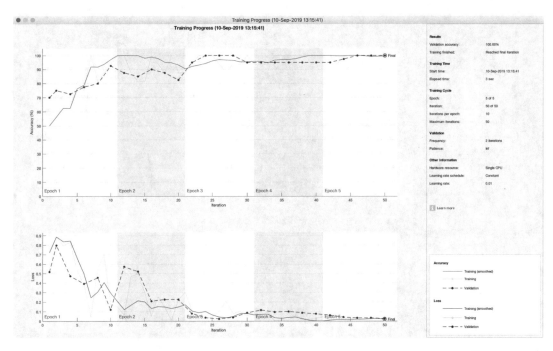

Figure 3.8: *The training window with a learn rate of 0.01 and a leaky reluLayer. The elapsed time is four seconds.*

```
    Show all properties

  Accuracy is    97.50%
```

We can try different activation functions. The script `EllipsesNeuralNetLeaky.m` shows a leaky reluLayer. We replaced `reluLayer` with `leakyReluLayer`. The output is similar, but in this case, learning was achieved even faster than before. See Figure 3.8 for a training run.

EllipsesNeuralNetLeaky.m

```
29  % This gives the structure of the convolutional neural net
30  layers = [
31      imageInputLayer(size(img))
32
33      convolution2dLayer(3,8,'Padding','same')
34      batchNormalizationLayer
35      leakyReluLayer
36
37      maxPooling2dLayer(2,'Stride',2)
38
39      convolution2dLayer(3,16,'Padding','same')
40      batchNormalizationLayer
41      leakyReluLayer
```

```
42
43        maxPooling2dLayer(2,'Stride',2)
44
45        convolution2dLayer(3,32,'Padding','same')
46        batchNormalizationLayer
47        leakyReluLayer
48
49        fullyConnectedLayer(2)
50        softmaxLayer
51        classificationLayer
52           ];
```

The output with the leaky layer is as follows:

```
>> EllipsesNeuralNetLeaky

ans =

  Figure (1: Ellipses) with properties:

        Number: 1
          Name: 'Ellipses'T
         Color: [0.9400 0.9400 0.9400]
      Position: [560 528 560 420]
         Units: 'pixels'

  Show all properties

Accuracy is    84.25%
```

We can try fewer layers. `EllipsesNeuralNetOneLayer.m` has only one set of layers.

EllipsesNeuralNetOneLayer.m

```
29  %% Define the layers for the net
30  % This gives the structure of the convolutional neural net
31  layers = [
32      imageInputLayer(size(img))
33
34      convolution2dLayer(3,8,'Padding','same')
35      batchNormalizationLayer
36      reluLayer
37
38      fullyConnectedLayer(2)
39      softmaxLayer
40      classificationLayer
41         ];
42
43  analyzeNetwork(layers)
```

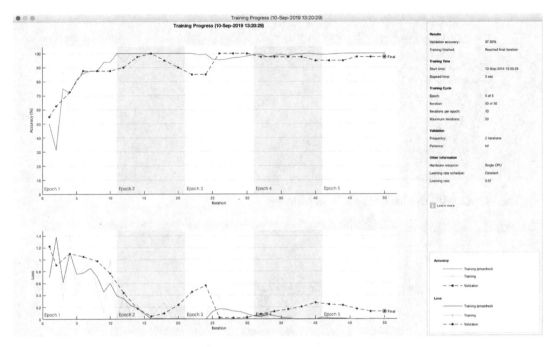

Figure 3.9: *The training window for a net with one set of layers.*

The training window is shown in Figure 3.9. The results with only one set of layers are not as good as the neural network with multiple layers. The accuracy is reduced but it is still acceptable. This shows that you need to try different options with your net architecture as well.

```
>> EllipsesNeuralNetOneLayer
ans =
  Figure (2: Ellipses) with properties:

       Number: 2
         Name: 'Ellipses'
        Color: [0.9400 0.9400 0.9400]
     Position: [560 528 560 420]
        Units: 'pixels'

  Show all properties
Accuracy is    87.25%
```

The one-set network is short enough that the whole thing can be visualized inside the window of `analyzeNetwork`, as in Figure 3.10. This function will check your layer architecture before you start training and alert you to any errors. The size of the activations and "Learnables" is displayed explicitly.

In this chapter, we both generated our own image data and trained a neural net to classify features in our images. In this example, we were able to achieve 100% accuracy, but not after some debugging was required in creating and naming the images. It is critical to carefully

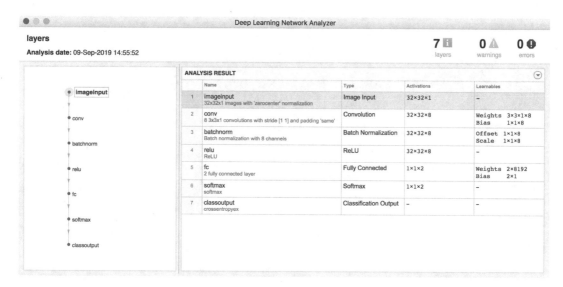

Figure 3.10: *The analyze window for the one-set convolutional network.*

examine your training and test data to ensure it contains the features you wish to identify. You should be prepared to experiment with your layers and training parameters as you develop nets for different problems.

CHAPTER 4

■ ■ ■

Classifying Movies

4.1   Introduction

Netflix, Hulu, and Amazon Prime all attempt to help you pick movies. In this chapter, we will create a database of movies, with fictional ratings. We will then create a set of viewers. We will then try to predict if a viewer would choose to watch a particular movie. We will use deep learning with MATLAB's pattern recognition network, `patternnet`. You will see that we can achieve accuracies of up to 100% over our small set of movies. Guessing what a customer would like to buy is something that all manufacturers and retailers want to do as it lets them focus their efforts on products that are of the greatest interest to their customers. As we show in this chapter, deep learning can be a valuable tool.

4.2   Generating a Movie Database

4.2.1 Problem

We first need to generate a database of movies.

4.2.2 Solution

Write a MATLAB function, `CreateMovieDatabase.m`, to create a database of movies. The movies will have fields for genre, reviewer ratings (like IMDB), and viewer ratings.

4.2.3 How It Works

We first need to come up with a method for characterizing movies. Table 4.1 gives our system. MPAA stands for Motion Picture Association of America. It is an organization that rates movies. Other systems are possible, but this will be sufficient to test out our deep learning system. Three of the data points that will be used are strings and two are numbers. One number, length, is a continuum, while ratings have discrete values. The second number, quality, is based on the "stars" in the rating. Some movie databases, like IMDB, have fractional values because they average over all their users. We created our own MPAA ratings and genres based on our opinions. The real MPAA ratings may be different.

© Michael Paluszek, Stephanie Thomas, Eric Ham 2022
M. Paluszek et al., *Practical MATLAB Deep Learning*,
https://doi.org/10.1007/978-1-4842-7912-0_4

Table 4.1: *The movie database will have five characteristics.*

Characteristic	Value
Name	String
Genre	Type of movie (Animated, Comedy, Dance, Drama, Fantasy, Romance, SciFi, War) in a string
Rating	Average viewer rating string, one to four stars, such as from IMBD
Quality	Number of stars
Length	Minute duration
MPAA Rating	MPAA Rating (PG, PG-13, R) in a string

Length can be any duration. We'll use `randn` to generate the lengths around a mean of 1.8 hours and a standard deviation of 0.15 hours. Length is a floating-point number. Stars are one to five and must be integers.

We created an Excel file with the names of 100 real movies, which is included with the book's software. We assigned random genres and MPAA ratings (PG, R, and so forth) to them. We then saved the Excel file as tab-delimited text and search for tabs in each line. (There are other ways to import data from Excel and text files in MATLAB; this is just one example.) We then assign the data to the fields. The function will check to see if the maximum length or rating is zero, which it is for all the movies in this case, and then create random values. You can create a spreadsheet with rating values as an extension of this recipe. We use `str2double` since it is faster than `str2num` when you know that the value is a single number. `fgetl` reads in one line and ignores the end of line characters.

You'll notice that we check for `NaN` in the length and rating fields since

```
>> str2double('')

ans =

    NaN
```

CreateMovieDatabase.m

```
21  function d = CreateMovieDatabase( file )
22
23  if( nargin < 1 )
24      Demo
25      return
26  end
27
28  f = fopen(file,'r');
29
30  d.name    = {};
31  d.rating  = [];
32  d.length  = [];
33  d.genre   = {};
34  d.mPAA    = {};
```

```
35  t              = sprintf('\t');   % a tab character
36  k              = 0;
37
38  while(~feof(f))
39    k                = k + 1;
40    q                = fgetl(f);       % one line of the file
41    j                = strfind(q,t);   % find the tabs in the line
42    d.name{k}        = q(1:j(1)-1);    % the name is the first token
43    d.rating(1,k)    = str2double(q(j(1)+1:j(2)-1));
44    d.genre{k}       = q(j(2)+1:j(3)-1);
45          d.length(1,k)    = str2double(q(j(3)+1:j(4)-1));
46    d.mPAA{k}        = q(j(4)+1:end);
47  end % end of the file
48
49  if( max(d.rating) == 0 || isnan(d.rating(1)) )
50    d.rating = randi(5,1,k);
51  end
52
53  if( max(d.length) == 0 || isnan(d.length(1)))
54    d.length = 1.8 + 0.15*randn(1,k);
55  end
56
57  fclose(f);
```

Running the function demo, shown in the following, creates a database of movies in a data structure. The length and ratings are added here.

```
60  function Demo
61
62  file = 'Movies.txt';
63  d    = CreateMovieDatabase( file )
```

The output is the following:

```
>> CreateMovieDatabase

d =

  struct with fields:

      name: {1 x 100 cell}
    rating: [1 x 100 double]
    length: [1 x 100 double]
     genre: {1 x 100 cell}
      mPAA: {1 x 100 cell}
```

Here are the first few movies:

```
>> d.name'
ans =
  100x1 cell array
```

```
{'2001: A Space Odyssey'              }
{'A Star is Born'                     }
{'Alien'                              }
{'Aliens'                             }
{'Amadeus'                            }
{'Apocalypse Now'                     }
{'Apollo 13'                          }
{'Back to the Future'                 }
```

4.3 Generating a Viewer Database

4.3.1 Problem

We next need to generate a database of movie viewers for training and testing.

4.3.2 Solution

Write a MATLAB function, `CreateMovieViewers.m`, to create a series of watchers. We will use a probability model to select which of the movies each viewer has watched based on the movie's genre, length, and ratings.

4.3.3 How It Works

Each watcher will have seen a fraction of the 100 movies in our database. This will be a random integer between 20 and 60. Each movie watcher will have a probability for each characteristic: the probability that they would watch a movie rated one or five stars, the probability that they would watch a movie in a given genre, etc. (Some viewers enjoy watching so-called "turkeys"!) We will combine the probabilities to determine the movies the viewer has watched. For `mPAA`, `genre`, and `rating`, the probabilities will be discrete. For the length, it will be a continuous distribution. You could argue that a watcher would always want the highest-rated movie, but remember this rating is based on an aggregate of other people's opinions and so may not directly map onto the particular viewer. The only output of this function is a list of movie numbers for each user. The list is in a cell array.

We start by creating cell arrays of the categories. We then loop through the viewers and compute probabilities for each movie characteristic. We then loop through the movies and compute the combined probabilities. This results in a list of movies watched by each viewer.

```
16  function [mvr,pWatched] = CreateMovieViewers( nViewers, d )
17
18  if( nargin < 1 )
19    Demo
20    return
21  end
22
23  mvr   = cell(1,nViewers);
24  nMov  = length(d.name);
25  genre = { 'Animated', 'Comedy', 'Dance', 'Drama', 'Fantasy', 'Romance'
            ,...
```

```
26              'SciFi', 'War', 'Horror', 'Music', 'Crime'};
27  mPAA  = {'PG-13','R','PG'};
28
29  % Loop through viewers. The inner loop is movies.
30  for j = 1:nViewers
31     % Probability of watching each MPAA
32     rMPAA = rand(1,length(mPAA));
33     rMPAA = rMPAA/sum(rMPAA);
34
35     % Probability of watching each Rating (1 to 5 stars)
36     r = rand(1,5);
37     r = r/sum(r);
38
39     % Probability of watching a given Length
40     mu    = 1.5 + 0.5*rand; % preferred movie length, between 1.5 and 2
              hrs
41     sigma = 0.5*rand;       % variance, up to 1/2 hour
42
43     % Probability of watching by Genre
44     rGenre = rand(1,length(genre));
45     rGenre = rGenre/sum(rGenre);
46
47     % Compute the likelihood the viewer watched each movie
48     pWatched = zeros(1,nMov);
49     for k = 1:nMov
50        pRating  = r(d.rating(k));           % probability for this rating
51        i        = strcmp(d.mPAA{k},mPAA);   % logical array with one
              match
52        pMPAA    = rMPAA(i);                 % probability for this MPAA
53        i        = strcmp(d.genre{k},genre); % logical array
54        pGenre   = rGenre(i);                % probability for this
              genre
55        pLength  = Gaussian(d.length(k),sigma,mu);  % probability for this
              length
56        pWatched(k) = 1 - (1-pRating)*(1-pMPAA)*(1-pGenre)*(1-pLength);
57     end
58
59     % Sort the movies and pick the most likely to have been watched
60     nInterval = floor( [0.2 0.6]*nMov );
61     nMovies = randi(nInterval);
62     [~,i]    = sort(pWatched);
63     mvr{j}   = i(1:nMovies);
64  end
```

This code computes the Gaussian or normal probability. The inputs include a standard deviation sigma and a mean mu.

```
66   function p = Gaussian(x,sigma,mu)
67   %% CreateMovieViewers>Gaussian
68   % The probability is 1 when x==mu and declines for shorter or longer
         movies
69
70   p = exp(-(x-mu)^2/(2*sigma^2));
```

The built-in function demo follows the Gaussian function. It is run automatically if the function is called with no inputs.

```
72   function Demo
73   %% CreateMovieViewers>Demo
74   % Load the Movies.mat database and create 4 viewers.
75
76   s = load('Movies.mat');
77
78   mvr = CreateMovieViewers( 4, s.d )
```

The output from the demo is as follows. This shows how many of the movies in the database each viewer has watched; the most is 57 movies and the fewest is 26.

```
1   >> CreateMovieViewers
2
3   mvr =
4
5       1x4 cell array
6
7         {1x33 double}    {1x57 double}    {1x51 double}    {1x26 double}
```

4.4   Training and Testing

4.4.1 Problem

We want to test a deep learning algorithm to select new movies for the viewer, based on what the algorithm thinks a viewer would choose to watch.

4.4.2 Solution

Create a viewer database and train a pattern recognition neural net on the viewer's movie selections. This is done in the script MovieNN.m. We will train a neural net for each viewer in the database.

4.4.3 How It Works

First, the movie data is loaded and displayed.

MovieNN.m

```
14  %% The movies
15  s = load('Movies.mat');
16  NewFigure('Movie Data')
17  subplot(2,2,1)
18  histogram(s.d.length)
19  xlabel('Movie Length')
20  ylabel('# Movies')
21  subplot(2,2,2)
22  histogram(s.d.rating)
23  xlabel('Stars')
24  ylabel('# Movies')
25  subplot(2,1,2)
26  histogram(categorical(s.d.genre))
27  ylabel('# Movies')
28  set(gca,'xticklabelrotation',90)
```

The viewer database is then created from the movie database:

MovieNN.m

```
30  %% The movie viewers
31  nViewers  = 4;
32  mvr       = CreateMovieViewers( nViewers, s.d );
```

The next block displays the characteristics of the movies each viewer has watched. This is shown graphically in Figure 4.1. For the moment, there are only four viewers.

```
34  % Display the movie viewer's data
35  lX        = linspace(min(s.d.length),max(s.d.length),5);
36
37  for k = 1:nViewers
38    NewFigure(sprintf('Viewer %d',k));
39
40    subplot(2,2,1);
41    g = zeros(1,11);
42    for j = 1:length(mvr{k})
43      i       = mvr{k}(j);
44      l       = strmatch(s.d.genre{i},genre); %#ok<MATCH2>
45      g(l)    = g(l) + 1;
46    end
47    bar(1:11,g);
48    set(gca,'xticklabel',genre,'xticklabelrotation',90,'xtick',1:11)
49    xlabel('Genre')
50    title(sprintf('Viewer %d',k))
51    grid on
52
```

```
53    subplot(2,2,2);
54    g = zeros(1,5);
55    for j = 1:5
56       for i = 1:length(mvr{k})
57          if( s.d.length(mvr{k}(i)) > lX(j) )
58             g(j)  = g(j) + 1;
59          end
60       end
61    end
62    bar(1:5,g);
63    set(gca,'xticklabel',floor(lX*60),'xtick',1:5)
64    xlabel('Length Greater Than (min)')
65    grid on
66
67    subplot(2,2,3);
68    g = zeros(1,5);
```

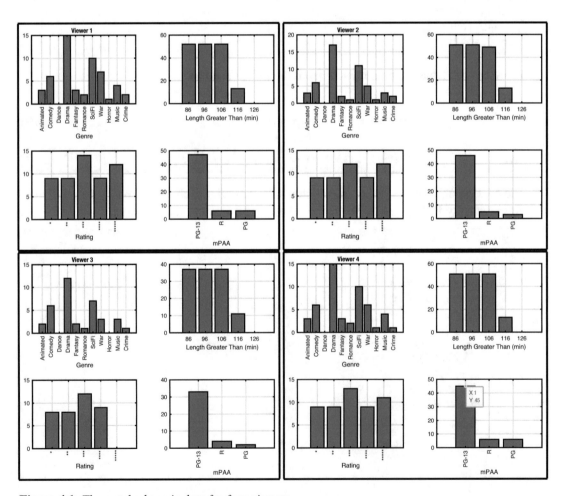

Figure 4.1: *The watched movie data for four viewers.*

```
69      for j = 1:length(mvr{k})
70          i    = mvr{k}(j);
71          l    = s.d.rating(i);
72         g(l)  = g(l) + 1;
73      end
74      bar(1:5,g);
75      set(gca,'xticklabel',rating,'xticklabelrotation',90,'xtick',1:5)
76      xlabel('Rating')
77      grid on
78
79      subplot(2,2,4);
80      g = zeros(1,3);
81      for j = 1:length(mvr{k})
82          i    = mvr{k}(j);
83          l    = strmatch(s.d.mPAA{i},mPAA); %#ok<MATCH2>
84         g(l)  = g(l) + 1;
85      end
86      bar(1:3,g);
87      set(gca,'xticklabel',mPAA,'xticklabelrotation',90,'xtick',1:3)
88      xlabel('mPAA')
89      grid on
90  end
```

We use bar charts throughout. Notice how we make the x labels strings for the genre and so on. We also rotate them 90 degrees for clarity. The length is the number of movies longer than the number on the x-axis.

This data is based on our viewer model from a recipe in Section 4.3 which is based on joint probabilities. We will train the neural net on a subset of the movies. This is a classification problem. We just want to know if a given movie would be picked or not picked by the viewer.

We use `patternnet` to predict the movies watched. This is shown in the next code block. The input to `patternnet` is the sizes of the hidden layers, in this case, a single layer of size 40. We convert everything into integers. Note that you need to round the results since `patternnet` does not return integers, despite the label being an integer. `patternnet` has methods `train` and `view`.

```
92  %% Train and test the neural net for each viewer
93  for k = 1:nViewers
94    % Create the training arrays
95    x = zeros(4,100); % the input data
96    y = zeros(1,100); % the target - did the viewer watch the movie?
97
98    nMov = length(mvr{k}); % number of watched movies
99    for j = 1:nMov
100       i       = mvr{k}(j); % index of the jth movie watched by the kth
                viewer
101      x(1,j) = s.d.rating(i);
102      x(2,j) = s.d.length(i);
103      x(3,j) = strmatch(s.d.mPAA{i},mPAA,'exact'); %#ok<*MATCH3>
104      x(4,j) = strmatch(s.d.genre{i},genre,'exact');
```

```
105     y(1,j) = 1; % movie watched
106   end
107
108   i = setdiff(1:100,mvr{k}); % unwatched movies
109   for j = 1:length(i)
110     x(1,nMov+j) = s.d.rating(i(j));
111     x(2,nMov+j) = s.d.length(i(j));
112     x(3,nMov+j) = strmatch(s.d.mPAA{i(j)},mPAA,'exact');
113     x(4,nMov+j) = strmatch(s.d.genre{i(j)},genre,'exact');
114     y(1,nMov+j) = 0; % movie not watched
115   end
116
117   % Create the training and testing data
118   j         = randperm(100);
119   j         = j(1:70);  % train using 70% of the available data
120   xTrain  = x(:,j);
121   yTrain  = y(1,j);
122   j         = setdiff(1:100,j);
123      xTest   = x(:,j);
124   yTest   = y(1,j);
125
126   net      = patternnet(40); % input a scalar or row of layer sizes
127   net      = train(net,xTrain,yTrain);
128   view(net);
129   yPred = round(net(xTest));
130
131   %% Test the neural net
132   accuracy = sum(yPred == yTest)/length(yTest);
133   fprintf('Accuracy for viewer %d (%d movies watched) is %8.2f%%\n',...
134      k,nMov,accuracy*100)
135   end
```

The training window is shown in Figure 4.3. When we view the net, MATLAB opens the display in Figure 4.2. Each net has four inputs, for the movie's rating, length, genre, and MPAA classification. The net's single output is the classification of whether the viewer has watched the movie or not. The training window provides access to additional plots of the training and performance data. We train with 70% of the data.

The output of the script for `patternnet(40)` is shown in the following. The accuracy is the percentage of the movies in the test set (30% of the data available) that the net correctly predicted the viewer has watched. The accuracy is usually between 65% and 94% for this size hidden layer.

```
>> MovieNN
Accuracy for viewer 1 (21 movies watched) is    93.33%
Accuracy for viewer 2 (52 movies watched) is    70.00%
Accuracy for viewer 3 (59 movies watched) is    66.67%
Accuracy for viewer 4 (50 movies watched) is    93.33%
```

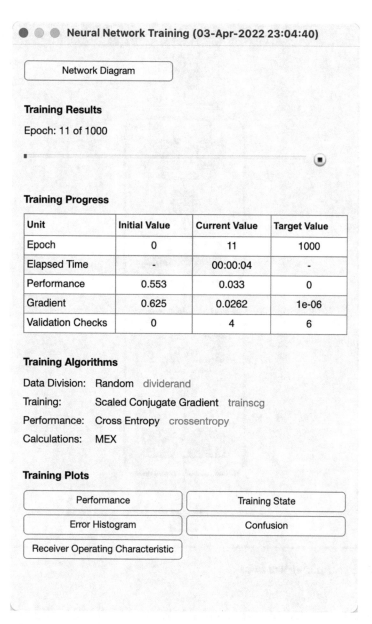

Figure 4.2: *The patternnet network with four inputs and one output.*

`patternnet(40)` returns good results, but we also tried greater and smaller numbers of layers and multiple hidden layers. For example, with a layer size of just 5, the accuracy ranges from 50 to 70%. With a size of 50, we reached over 90% of all viewers! Granted, this is a small number of movies. The results will vary with each run due to the random nature of the

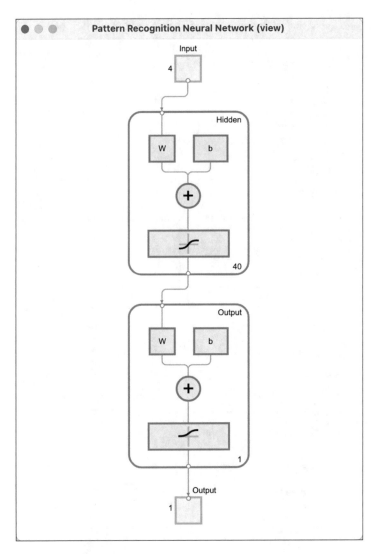

Figure 4.3: *Patternnet training window.*

variables in the test. The predictions are probably as good as Netflix! It is important to note that the neural network did not know anything about the viewer model. Nonetheless, it does a good job of predicting movies that the viewer might like.

CHAPTER 5

■ ■ ■

Algorithmic Deep Learning

In this chapter, we introduce the Algorithmic Deep Learning Neural Network (ADLNN), a deep learning system that incorporates algorithmic descriptions of the processes as part of the deep learning neural network. The dynamical models provide domain knowledge. These are in the form of differential equations. The outputs of the network are both indications of failures and updates to the parameters of the models. Training can be done using simulations, prior to operations, or through operator interaction during operations.

The system is shown in Figure 5.1. This is based on work from the books by Paluszek and Thomas [41, 42]. These books show the relationships between machine learning, adaptive control, and estimation. This model can be encapsulated in a set of differential equations. We will limit ourselves to sensor failures in this example. The output indicates what kind of failures have occurred. It indicates that either one or both of the sensors have failed.

Figure 5.2 shows an air turbine [26]. This air turbine has a constant pressure air supply. The pressurized air causes the turbine to spin. It is a way to produce rotary motion for a drill or other purposes.

We can control the valve from the air supply, the pressure regulator, to control the speed of the turbine. The air flows past the turbine blades causing it to turn. The control needs to adjust the air pressure to handle variations in the load. The load is the resistance to turning. For example, a drill might hit a harder material while in use. We measure the air pressure p downstream from the valve, and we also measure the rotational speed of the turbine ω with a tachometer.

The state vector is

$$\begin{bmatrix} p \\ \omega \end{bmatrix} \tag{5.1}$$

where ω is the tachometer rate of the turbine and p is the pressure.

The dynamical model for the air turbine is

$$\begin{bmatrix} \dot{p} \\ \dot{\omega} \end{bmatrix} = \begin{bmatrix} -\frac{1}{\tau_p} & 0 \\ \frac{K_t}{\tau_t} & -\frac{1}{\tau_t} \end{bmatrix} \begin{bmatrix} p \\ \omega \end{bmatrix} + \begin{bmatrix} \frac{K_p}{\tau_p} \\ 0 \end{bmatrix} u \tag{5.2}$$

© Michael Paluszek, Stephanie Thomas, Eric Ham 2022
M. Paluszek et al., *Practical MATLAB Deep Learning*,
https://doi.org/10.1007/978-1-4842-7912-0_5

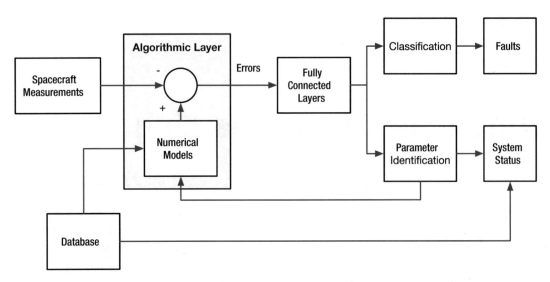

Figure 5.1: *Algorithmic Deep Learning Neural Network (ADLNN). The network uses numerical models as a filtering layer. The numerical models are a set of differential equations configured as a detection filter.*

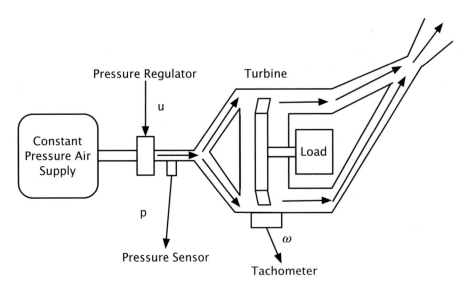

Figure 5.2: *Air turbine. The arrows show the airflow. The air flows through the turbine blade tips causing it to turn.*

This is a state space system

$$\dot{x} = ax + bu \tag{5.3}$$

where

$$a = \begin{bmatrix} -\frac{1}{\tau_p} & 0 \\ \frac{K_t}{\tau_t} & -\frac{1}{\tau_t} \end{bmatrix} \tag{5.4}$$

$$b = \begin{bmatrix} \frac{K_p}{\tau_p} \\ 0 \end{bmatrix} \tag{5.5}$$

The pressure downstream from the regulator is equal to $K_p u$ when the system is in equilibrium. τ_p is the regulator time constant, and τ_t is the turbine time constant. The turbine speed is $K_t p$ when the system is in equilibrium. The tachometer measures ω, and the pressure sensor measures p. The load is folded into the time constant for the turbine.

The code for the right-hand side of the dynamical equations is shown in the following. Only one line of code, line 44 in RHSAirTurbine.m, is the right-hand side. The rest returns the default data structure. The simplicity of the model is due to its being a state space model. The number of states could be large, yet the code would not change. As you can see, the dynamical equations are just one line of code.

RHSAirTurbine.m

```
1   %% RHSAIRTURBINE Air turbine dynamics model.
25
26  % Default data structure
27  if( nargin < 1 )
28     kP    = 1;
29     kT    = 2;
30     tauP  = 10;
31     tauT  = 40;
32     c     = eye(2);
33     b     = [kP/tauP;0];
34     a     = [-1/tauP 0; kT/tauT -1/tauT];
35
36     xDot = struct('a',a,'b',b,'c',c,'u',0);
37     if( nargout == 0)
38        disp('RHSAirTurbine struct:');
39     end
40     return
41  end
42
43  % Derivative
44  xDot = d.a*x + d.b*d.u;
```

The simulation, AirTurbineSim.m, is shown in the following. The control is a constant, also known as a step input. TimeLabel converts the time vector into units (minutes, hours, etc.) that are easier to read. It also returns a label for the time units that you can use in the plots.

AirTurbineSim.m

```
1   %% Script with simulation of an air turbine
2   % Simulates an air turbine.
3   %% See also
4   % RHSAirTurbine, RungeKutta, TimeLabel, PlotSet
5
6   %% Initialization
7   tEnd    = 1000; % sec
8
9   % State space system
10  d       = RHSAirTurbine;
11
12  % This is the regulator input.
13  d.u     = 100;
14
15  dT      = 0.02; % sec
16  n       = ceil(tEnd/dT);
17
18  % Initial state
19  x       = [0;0];
20
21  %% Run the simulation
22
23  % Plotting array
24  xP      = zeros(2,n);
25  t       = (0:n-1)*dT;
26
27  for k = 1:n
28    xP(:,k) = x;
29    x       = RungeKutta( @RHSAirTurbine, t(k), x, dT, d );
30  end
31
32  %% Plot the states and residuals
33  [t,tL] = TimeLabel(t);
34  yL      = {'p (N/m^2)' '\omega (rad/s)' };
35  tTL     = 'Air Turbine Simulation';
36  PlotSet( t, xP,'x label',tL,'y label',yL,'figure title',tTL)
```

The response to a step input for u is shown in Figure 5.3. The pressure settles faster than the turbine angular velocity. This is due to the turbine time constant and the lag in the pressure change.

Now that we understand better how an air turbine works, we can build the filter to detect its sensor failures. It is always a good idea to understand your dynamical system. When building an algorithmic filter or estimator, this is a necessity. For a neural net, it is not necessary just to get something working, but really helps when interpreting the neural net performance.

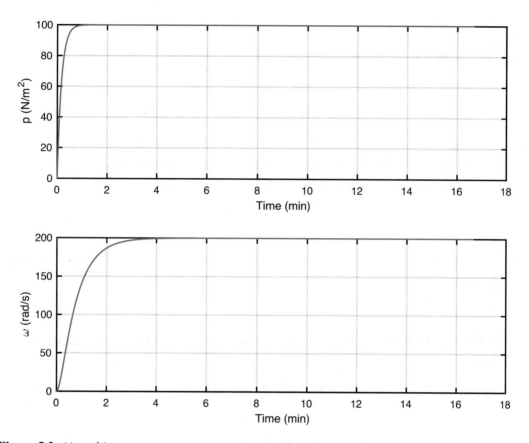

Figure 5.3: *Air turbine response to a step pressure regulator input.*

5.1   Building the Filter

5.1.1 Problem

We want to build a system to detect regulator and sensor failures in our air turbine using the linear model developed in the previous recipe.

5.1.2 Solution

We will build a detection filter that detects pressure regulator failures and tachometer failures. Our plant model (continuous a, b, and c state space matrices) will be an input to the filter building function.

5.1.3 How It Works

The detection filter is an estimator with a specific gain matrix that multiplies the residuals.

$$
\begin{bmatrix} \dot{\hat{p}} \\ \dot{\hat{\omega}} \end{bmatrix} = \begin{bmatrix} -\frac{1}{\tau_p} & 0 \\ \frac{K_t}{\tau_t} & -\frac{1}{\tau_t} \end{bmatrix} \begin{bmatrix} \hat{p} \\ \hat{\omega} \end{bmatrix} + \begin{bmatrix} \frac{K_p}{\tau_p} \\ 0 \end{bmatrix} u + \begin{bmatrix} d_{11} & d_{12} \\ d_{21} & d_{22} \end{bmatrix} \begin{bmatrix} p - \hat{p} \\ \omega - \hat{\omega} \end{bmatrix} \tag{5.6}
$$

where $\hat{p}$ is the estimated pressure and $\hat{\omega}$ is the estimated angular rate of the turbine. The D matrix is the matrix of detection filter gains. This matrix multiplies the residuals, the difference between the measured and estimated states, into the detection filter. The residual vector is

$$
r = \begin{bmatrix} p - \hat{p} \\ \omega - \hat{\omega} \end{bmatrix} \tag{5.7}
$$

The D matrix needs to be selected so that this vector tells us the nature of the failure. The gains should be selected so that

1. The filter is stable.

2. If the pressure regulator fails, the first residual, $p - \hat{p}$, is nonzero, but the second remains zero.

3. If the turbine fails, the second residual $\omega - \hat{\omega}$ is nonzero, but the first remains zero.

The gain matrix is

$$
D = a + \begin{bmatrix} \frac{1}{\tau_1} & 0 \\ 0 & \frac{1}{\tau_2} \end{bmatrix} \tag{5.8}
$$

We can see this by substituting this D into Equation 5.6:

$$
\begin{bmatrix} \dot{\hat{p}} \\ \dot{\hat{\omega}} \end{bmatrix} = \begin{bmatrix} a_{11} & a_{12} \\ a_{21} + \frac{K_t}{\tau_t} & a_{22} \end{bmatrix} \begin{bmatrix} \hat{p} \\ \hat{\omega} \end{bmatrix} + \begin{bmatrix} \frac{K_p}{\tau_p} \\ 0 \end{bmatrix} u + D \begin{bmatrix} p \\ \omega \end{bmatrix} \tag{5.9}
$$

The time constant τ_1 is the pressure residual time constant. The time constant τ_2 is the tachometer residual time constant. In effect, we cancel out the dynamics of the plant and replace them with decoupled detection filter dynamics. These time constants should be shorter than the time constants in the dynamical model so that we detect failures quickly. However, they need to be at least twice as long as the sampling period to prevent numerical instabilities.

We will write a function, DetectionFilter, with three actions: an initialize case, an update case, and a reset case. varargin is used to allow the three cases to have different input lists. The function signature is

```
46  function d = DetectionFilter( action, varargin )
```

It can be called in three ways:

```
>> d = DetectionFilter( 'initialize', d, tau, dT )
>> d = DetectionFilter( 'update', u, y, d )
>> d = DetectionFilter( 'reset', d )
```

The first initializes the function, the second is called each time step for an update, and the last resets the filter. All data is stored in the data structure d.

The function simulates detecting failures of an air turbine. An air turbine has a constant pressure air source that sends air through a duct that drives the turbine blades. The turbine is attached to a load. The air turbine model is linear. Failures are modeled by multiplying the regulator input and tachometer output by a constant. A constant of zero is a total failure, and one is perfect operation.

The filter is built and initialized in the following code in DetectionFilter.m. The continuous state space model of the plant, in this case, our linear air turbine model, is an input. The selected time constants τ are also an input, and they are added to the plant model as in Equation 5.8. The function discretizes the plant a and b matrices and the computed detection filter gain matrix d.

DetectionFilter.m

```
49    case 'initialize'
50        d   = varargin{1};
51        tau = varargin{2};
52        dT  = varargin{3};
53
54        % Design the detection filter
55        d.d = d.a + diag(1./tau);
56
57        % Discretize both
58        d.d          = CToDZOH( d.d, d.b, dT );
59        [d.a, d.b] = CToDZOH( d.a, d.b, dT );
60
61        % Initialize the state
62        m   = size(d.a,1);
63        d.x = zeros(m,1);
64        d.r = zeros(m,1);
```

The update for the detection filter is in the same function. Note the equations implemented as described in the header.

```
66    case 'update'
67        u   = varargin{1};
68        y   = varargin{2};
69        d   = varargin{3};
70        r   = y - d.c*d.x;
71        d.x = d.a*d.x + d.b*u + d.d*r;
72        d.r = r;
```

Finally, we create a reset action to allow us to reset the residual and state values for the filter in between simulations.

```
74    case 'reset'
75        d   = varargin{1};
76        m   = size(d.a,1);
77        d.x = zeros(m,1);
78        d.r = zeros(m,1);
```

5.2 Simulating

5.2.1 Problem

We want to simulate a failure in the plant and demonstrate the performance of the failure detection.

5.2.2 Solution

We will build a MATLAB script that designs the detection filter using the function from the previous recipe and then simulates it with a user selectable pressure regulator or tachometer failure. The failure can be total or partial.

5.2.3 How It Works

The DetectionFilterSim.m script designs a detection filter using the Detection Filter function from the previous recipe and implements it in a loop. A Runge-Kutta integration propagates the continuous domain right-hand side of the air turbine, RHSAirTurbine. The detection filter is discrete time.

The script has two scale factors uF and tachF that multiply the regulator input and the tachometer output to simulate failures. Setting a scale factor to zero is a total failure, and setting it to one indicates that the device is working perfectly. If we fail one, we expect the associated residual to be nonzero and the other to stay at zero.

DetectionFilterSim.m

```
1  %% Script to simulate a detection filter
2  % Simulates detecting failures of an air turbine. An air turbine has a
     constant
3  % pressure air source that sends air through a duct that drives the
     turbine
4  % blades. The turbine is attached to a load.
5  %
6  % The air turbine model is linear. Failures are modeled by multiplying
     the
7  % regulator input and tachometer output by a constant. A constant of 0
     is a
8  % total failure and 1 is perfect operation.
9  %% See also:
```

```
14
15  % Time constants for failure detection
16  tau1 = 0.3; % sec
17  tau2 = 0.3; % sec
18
19  % End time
20  tEnd = 0.05; % sec
21
22  % State space system
23  d = RHSAirTurbine;
24
25  %% Initialization
26  dT = 0.001; % sec
27  n  = ceil(tEnd/dT);
28
29  % Initial state
30  x = [0;0];
31
32  %% Detection Filter design
33  dF = DetectionFilter('initialize',d,[tau1;tau2],dT);
34
35  %% Run the simulation
36
37  % Control. This is the regulator input.
38  u = 100;
39
40  % Plotting array
41  xP = zeros(4,n);
42  t  = (0:n-1)*dT;
43
44  for k = 1:n
45      % Measurement vector including measurement failure
46      y       = [x(1);tachF*x(2)]; % Sensor failure
47      xP(:,k) = [x;dF.r];
48
49      % Update the detection filter
50      dF      = DetectionFilter('update',u,y,dF);
51
52      % Integrate one step
53      d.u     = uF*u; % Actuator failure
54      x       = RungeKutta( @RHSAirTurbine, t(k), x, dT, d );
55  end
56
57  %% Plot the states and residuals
58  [t,tL] = TimeLabel(t);
59  yL      = {'p' '\omega' 'Residual P' 'Residual \omega' };
60  tTL     = 'Detection Filter Simulation';
61  PlotSet( t, xP,'x label',tL,'y label',yL,'figure title',tTL)
```

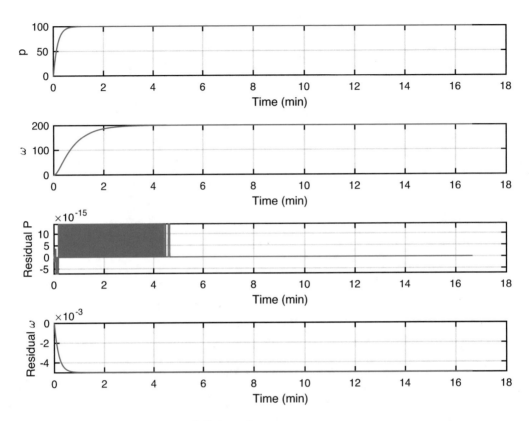

Figure 5.4: *Air turbine response to a failed regulator.*

In Figure 5.4, the regulator fails and its residual is nonzero. In Figure 5.5, the tachometer fails and its residual is nonzero. The residuals show what has failed clearly. Simple boolean logic (i.e., if-end statements) are all that is needed. Now, this indicates we don't need machine learning for this problem. The goal here is to show that machine learning can recognize the faults. This will demonstrate that it has the potential to be coupled with more complex systems. For the failed regulator, the time step and end time were reduced.

A detection filter is a type of filter. It filters out nonfailures, much like a low-pass filter filters out noise. Adding any type of filter stage to a deep learning system can enhance its performance. Of course, as with any filter, one needs to be careful not to filter out information needed by the learning system. For example, suppose you had an oscillator that was your dynamical system. If a low-pass filter cutoff were below the oscillation frequency, you would not be able to learn anything about the oscillation.

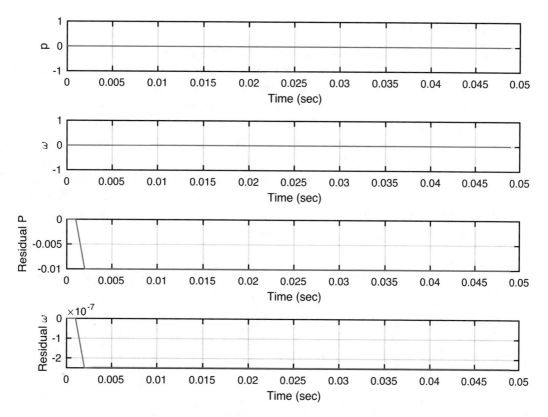

Figure 5.5: *Air turbine response to a failed tachometer. The residuals immediately reach the outputs indicating the failure because the filter is fast.*

5.3 Testing and Training

5.3.1 Problem

We want a neural network to characterize faults. Both tachometer and regulator failures will be characterized.

5.3.2 Solution

We use the same approach as we did for the XOR problem in Chapter 2. The outputs from the detection filter are classified. This could be done by simple boolean logic. The point is to show that a neural net can solve the same problem.

5.3.3 How It Works

We run the simulation to get all possible residuals and combine them in a residual two-by-four array. Our outputs are strings for the four cases. You can use strings instead of numbers as the classifier labels, which is better than using integers and then converting them to strings. This just makes the code cleaner and easier to understand. Someone else working on your code is less likely to misinterpret the outputs. We use `feedforwardnet` to implement the neural net. It has two layers, two inputs, and one output. The output is the status of the system. We first train the system with 600 randomly selected test cases. We then simulate the network.

The net is a `feedforwardnet` with two layers. There is one output, the failure, and two inputs from the detection filter. We measure the residuals that we expect to see for each possible failure case. There are four, "none," "both," "tach," and "regulator." The training pairs are a random selection from the four possible sets using `randi`.

DetectionFilterNN.m

```
4   % feedforwardnet, configure, train, sim
5
6   % Train the neural net
7   % Cases
8   % 2 layers
9   % 2 inputs
10  % 1 output
11
12  net       = feedforwardnet(2);
13
14  %             [none both         tach         regulator]
15  residual = [0        0.18693851  0            -0.18693851;...
16              0       -0.00008143 -0.09353033  -0.00008143];
17
18  % labels is a strings array
19  label     = ["none" "both"  "tach" "regulator"];
20
21  % How many sets of inputs
22  n    = 600;
23
24  % This determines the number of inputs and outputs
25  x    = zeros(2,n);
26  y    = zeros(1,n);
27
28  % Create training pairs
29  for k = 1:n
30    j       = randi([1,4]);
31    x(:,k)  = residual(:,j);
32    y(k)    = label(j);
33  end
34
```

```
35  net        = configure(net, x, y);
36  net.name   = 'DetectionFilter';
37  net        = train(net,x,y);
38  c          = sim(net,residual);
39
40  fprintf('\nRegulator  Tachometer    Failed\n');
41  for k = 1:4
42    fprintf('%9.2e %9.2e       %s\n',residual(1,k),residual(2,k),label(k))
        ;
43  end
44
45  % This only works for feedforwardnet(2);
46  fprintf('\nHidden layer biases %6.3f %6.3f\n',net.b{1});
47  fprintf('Output layer bias   %6.3f\n',net.b{2});
48  fprintf('Input layer weights  %6.2f %6.2f\n',net.IW{1}(1,:));
49  fprintf('                     %6.2f %6.2f\n',net.IW{1}(2,:));
50  fprintf('Output layer weights %6.2f %6.2f\n',net.LW{2,1}(1,:));
```

The training GUI is shown in Figure 5.6.

The GUI buttons are described in detail in Chapter 2 for the XOR problem.

The results are shown in the following. The neural net works quite well. The printout in the command window shows it uses two types of activation functions. The output layers use a linear activation function.

```
>> DetectionFilterNN

Regulator  Tachometer    Failed
 0.00e+00   0.00e+00     none
 1.87e-01  -8.14e-05     both
 0.00e+00  -9.35e-02     tach
-1.87e-01  -8.14e-05     regulator

Hidden layer biases -1.980  1.980
Output layer bias    0.159
Input layer weights      0.43   -1.93
                         0.65   -1.87
Output layer weights  -0.65    0.71
Hidden layer activation function tansig
Output layer activation function purelin
```

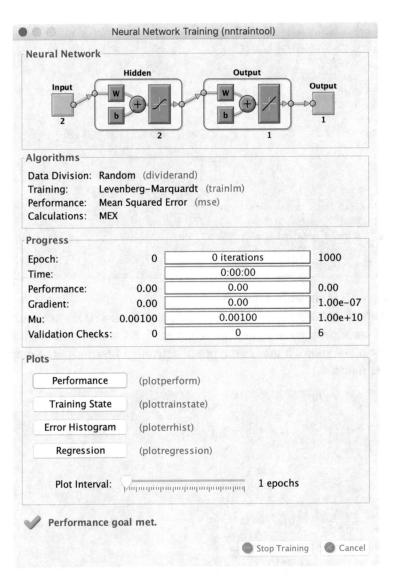

Figure 5.6: *The training GUI.*

CHAPTER 6

■ ■ ■

Tokamak Disruption Detection

6.1 Introduction

Tokamaks are fusion machines that are under development to produce baseload power. Baseload power is the power that is produced 24/7 and provides the base for powering the electric grid. The International Thermonuclear Experimental Reactor (ITER) is an international project that will produce net power from a Tokamak. Net power means the Tokamak produces more energy than it consumes. Consumption includes heating the plasma, controlling it, and powering all the auxiliary systems needed to maintain the plasma. It will allow researchers to study the physics of the Tokamak which will hopefully lead the way toward operational machines. A Tokamak is shown in Figure 6.1. The central solenoid field coils act like a transformer to initiate a plasma current. The outer poloidal and toroidal coils maintain the plasma. The plasma current itself produces its own magnetic field and induces currents in the other coils.

The image in Figure 6.1 was produced by the function `DrawTokamak` which calls `DCoil` and `SquareHoop`. We aren't going to discuss those three functions here. You should feel free to look through the functions as they show how easy it is to do 3D models using MATLAB.

One problem with Tokamaks is disruptions. A disruption is a massive loss of plasma control that extinguishes the plasma and results in large thermal and structural loads on the Tokamak wall. This can lead to catastrophic wall damage. This would be bad in an experimental machine and unacceptable in a power plant as it could lead to months of repairs.

The factors that can be used to predict a disruption [31] are

1. The poloidal beta (beta is the ratio of plasma pressure to magnetic pressure).

2. The line-integrated plasma density.

3. The plasma elongation.

4. The plasma volume divided by the device minor radius.

5. The plasma current.

6. The plasma internal inductance.

© Michael Paluszek, Stephanie Thomas, Eric Ham 2022   89
M. Paluszek et al., *Practical MATLAB Deep Learning*,
https://doi.org/10.1007/978-1-4842-7912-0_6

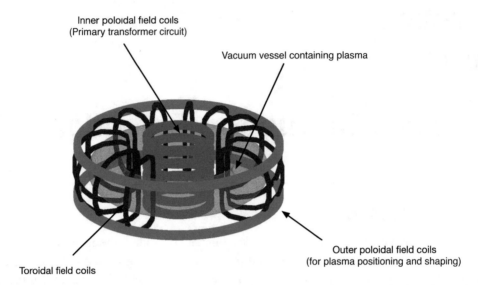

Inner poloidal field coils
(Primary transformer circuit)

Vacuum vessel containing plasma

Outer poloidal field coils
(for plasma positioning and shaping)

Toroidal field coils

Figure 6.1: *A Tokamak. There are three sets of coils. The central solenoid field coil initializes the plasma, and the poloidal and toroidal coils maintain the plasma. Some of the toroidal coils are left out to make it easier to see the Tokamak center.*

7. The locked mode amplitude.

8. The plasma vertical centroid position.

9. The total input power.

10. The safety factor reaching 95%. The safety factor is the ratio of the times a magnetic field line travels toroidally (the long way around the doughnut) vs. poloidally (the short way). We want the safety factor to be greater than one.

11. The total radiated power.

12. The time derivative of the stored diamagnetic energy, which is energy stored by the magnetic fields of the plasma.

Locked modes are magnetohydrodynamic (MHD) instabilities that are locked in phase and the laboratory frame. They can be precursors to disruptions. The plasma internal inductance is the inductance measured by integrating the inductance over the entire plasma. In a Tokamak, the poloidal direction is along the minor radius circumference. The toroidal direction is along the major radius circumference. In a plasma, the dipole moment due to the circulating current is in the opposite direction of the magnetic field which makes it diamagnetic. Diamagnetic energy is the energy stored in a magnetized plasma. Diamagnetic measurements measured this energy.

Our system is shown in Figure 6.2. We will just be looking at the plasma vertical position and the coil currents. We'll find it to be more than complex enough! We will start with the dynamics of the vertical motion of a plasma. We'll then learn about plasma disturbances. After that, we will design a vertical position controller. Finally, we'll get to deep learning.

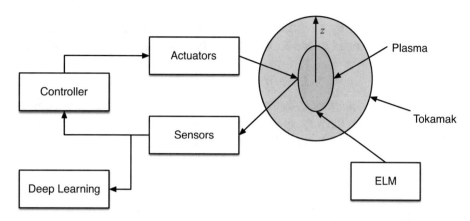

Figure 6.2: *Tokamak and control system. ELM means "Edge Localized Mode" and is a disturbance. The plasma is shown in a poloidal cut through the torus.*

6.2 Numerical Model

6.2.1 Dynamics

For our example, we need a numerical model of disruptions [8], [7], [37]. Ideally, our model would include all of the effects in the list given earlier. We use the model in Scibile [38]. We will only consider vertical movement.

The equilibrium force on the plasma is induced by the magnetic field and current density in the plasma:

$$J \times B = \nabla p \qquad (6.1)$$

where J is the current density, B is the magnetic field, and p is the pressure. J, B, and p are all three-element vectors, and that ∇ is a cross-product operation. Pressure is the force per unit area on the plasma. The momentum balance is [1]

$$\rho \frac{dv}{dt} = J \times B - \nabla p \qquad (6.2)$$

where v is the plasma velocity, and ρ is the plasma density. The imbalance causes plasma motion, that is, when $J \times B \neq \nabla p$. If we neglect the plasma mass, we get

$$L_P^T I + A_{PP} z I_p = F_p \qquad (6.3)$$

where L_p is the mutual change inductance matrix of the coils. I is the vector of currents in the Tokamak coils and the conducting shell around the plasma. F_p is the external force normalized to the plasma current I_p, and A_{PP} is the normalized destabilizing force. If we lump currents into active currents, driven by an external voltage, and passive currents, we get a simplified model of the plasma. We need to add Kirchhoff's voltage law to get a dynamical model.

$$L\dot{I} + RI + L_P\dot{z}I_p = \Gamma V \qquad (6.4)$$

Γ couples voltages to the currents, L is the coil inductance matrix, and R is the coil resistance. If we combine these, we get the state space matrices shown in the following. The lumped model is shown in Figure 6.3. Table 6.1 gives the model parameters.

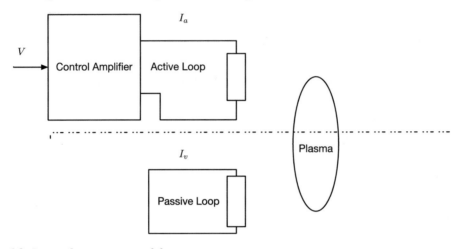

Figure 6.3: *Lumped parameter model.*

Table 6.1: *Model parameters from the Joint European Torus (JET). N/A is Newton (unit of force) per amp, in this case. Ω is Ohm, the unit of resistance; H is Henry, a unit of inductance; A is amps.*

Parameter	Description	JET	Units
L_{AA}	Active coil self-inductance	42.5×10^{-3}	H
$L_{AV} = L_{VA}$	Passive coil self-inductance	0.432×10^{-3}	H
L_{VV}	Active-passive coil mutual inductance	0.012×10^{-3}	H
R_{AA}	Active coil resistance	35.0×10^{-3}	Ω
R_{VV}	Passive coil resistance	2.56×10^{-3}	Ω
L'_{AP}	Mutual change inductance between the active coils and plasma displacement	115.2×10^{-6}	H/m
L'_{VP}	Mutual change inductance between the passive coils and plasma displacement	3.2×10^{-6}	H/m
A_{PP}	Normalized destabilizing force	0.5×10^{-6}	H/m²
F_P	Disturbance force normalized to the plasma current I_p	See ELM	N/A
τ_t	Controller lag	310×10^{-6}	s
I_p	Plasma current	1.5×10^6	A

dynamical equations are

$$
\begin{bmatrix} \dot{I}_a \\ \dot{I}_v \\ \dot{V}_a \end{bmatrix} = A^s \begin{bmatrix} I_a \\ I_v \\ V_a \end{bmatrix} + B^s \begin{bmatrix} V_c \\ \dot{F}_p \\ F_p \end{bmatrix}
\tag{6.5}
$$

$$
z = C^s \begin{bmatrix} I_a \\ I_v \\ V_a \end{bmatrix} + D^s \begin{bmatrix} V_c \\ \dot{F}_p \\ F_p \end{bmatrix}
\tag{6.6}
$$

$$
A^s = \frac{1}{1 - k_{av}} \begin{bmatrix} -\frac{R_{aa}}{L_{aa}} & k_{av}\frac{R_{vv}}{L_{av}} & L_{aa} \\ k_{av}\frac{R_{aa}}{L_{av}} & -\frac{R_{vv}}{L_{vv}}\frac{k_{av}-M_{vp}}{1-M_{vp}} & -L_{av} \\ 0 & 0 & -\frac{1-k_{av}}{\tau_t} \end{bmatrix}
\tag{6.7}
$$

$$
B^s = \begin{bmatrix} 0 & 0 & 0 \\ 0 & \frac{1}{L'_{vp}(1-M_{vp})} & 0 \\ \frac{1}{\tau_t} & 0 & 0 \end{bmatrix}
\tag{6.8}
$$

$$
C^s = \frac{1}{A''_{pp}I_p} \begin{bmatrix} -L'_{ap} & L'_{vp} & 0 \end{bmatrix}
\tag{6.9}
$$

$$
D^s = \begin{bmatrix} 0 & 0 & \frac{1}{A''_{pp}I_p} \end{bmatrix}
\tag{6.10}
$$

$$
k_{av} = \frac{L_{av}^2}{L_{aa}L_{vv}}
\tag{6.11}
$$

$$
M_{vp} = \frac{A''_{pp}L_{vv}}{L'^2_{vp}}
\tag{6.12}
$$

$$
M_{ap} = \frac{A''_{pp}L_{aa}}{-L'^2_{ap}}
\tag{6.13}
$$

This includes a first-order lag to replace the pure delay in Scibile. The preceding plasma dynamical equations may look very mysterious, but they are just a variation on a circuit with an inductor and a resistor, shown in Figure 6.4.

The major addition is that the currents produce forces that move the plasma in z.

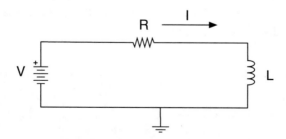

Figure 6.4: *Series resistor and inductor circuit, an RL circuit.*

The equation for this circuit is

$$L\frac{dI}{dt} + RI = V \tag{6.14}$$

which looks the same as our first-order lag. The first term is the voltage drop across the inductor and the second the voltage drop across the resistor. If we suddenly apply a constant V, we get the analytical solution:

$$I = \frac{V}{R}\left(1 - e^{-\frac{R}{L}t}\right) \tag{6.15}$$

L/R is the circuit's time constant τ. As $t \to \infty$, we get the equation for a resistor $V = IR$. You will notice a lot of R/L's in Equation 6.7.

6.2.2 Sensors

We are going to assume that we can measure the vertical position and the two currents directly. This is not entirely the case in a real machine. The vertical position is measured indirectly in a real machine. We also assume we have available the control voltage.

6.2.3 Disturbances

The disturbances are due to Edge Localized Modes (ELM). An Edge Localized Mode is a disruptive magnetohydrodynamic instability that occurs along the edges of a Tokamak plasma due to steep plasma pressure gradients [23]. The strong pressure gradient is called the edge pedestal. The edge pedestal improves plasma confinement time by a factor of two over the low-confinement mode. This is now the preferred mode of operation for Tokamaks. A simple model for an ELM is

$$d = k\left(e^{-\frac{t}{\tau_1}} - e^{-\frac{t}{\tau_2}}\right) \tag{6.16}$$

d is the output of the ELM. It can be scaled by k based on the usage. For example, in [38] it is scaled to show the output of a sensor. In our simulation, it is scaled to produce a driving force on the plasma.

$\tau_1 > \tau_2$ with the ELMs appearing randomly. The function ELM produces one ELM. The simulation must call it with a new sequence of times to get a new ELM. Figure 6.5 shows the results of the built-in demo in ELM. The function also computes the derivative since the derivative of the disturbance is also an input.

ELM.m

```
20   function eLM = ELM( tau1, tau2, k, t )
21
22   % Constants from the reference
23   if( nargin < 3 )
24     tau1 = 6.0e-4;
25     tau2 = 1.7e-4;
```

```
26    k     = 6.5;
27  end
28
29  % Reproduce the reference results
30  if( nargin < 4 )
31    t = linspace(0,12e-3);
32  end
33
34  d = k*[ exp( -t/tau1 ) - exp( -t/tau2 );...
35          exp( -t/tau2 )/tau2 - exp( -t/tau1 )/tau1 ];
36
37  if( nargout == 0 )
38    PlotSet( t*1000, d, 'x label', 'Time (ms)', 'y label', {'d' 'dd/dt'},
            'figure title', 'ELM' )
39  else
40    eLM = d;
41  end
```

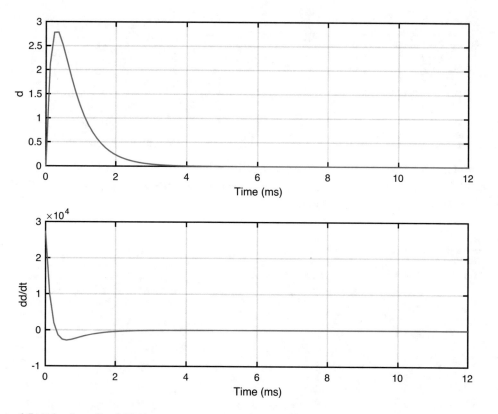

Figure 6.5: *Edge Localized Mode.*

6.2.4 Controller

We will use a controller to control the vertical position of the plasma, which otherwise is unstable as shown earlier. The controller will be a state space system using full state feedback. The states are the two currents. Position, z, is controlled indirectly. We will use a quadratic regulator. We will use a continuous version. This just means that we need to sample much faster than the range of frequencies for the control.

QCR.m

```
1    %% QCR Creates a regulator from a state space system.
23
24   function k = QCR( a, b, q, r )
25
26   if( nargin < 1 )
27     Demo
28     return
29   end
30
31   bor = b/r;
32
33   [sinf,rr] = Riccati( [a,-bor*b';-q',-a'] );
34
35   if( rr == 1 )
36     disp('Repeated roots. Adjust q or r');
37   end
38
39   k = r\(b'*sinf);
```

If you get repeated roots, you must manually adjust q or r, the state and control weights, respectively. The Matrix Riccati equation is solved in the subfunction Riccati. Notice the use of unique to find repeated roots.

QCR.m

```
41   %% QCR>Matrix Riccati equation
42   function [sinf, rr] = Riccati( g )
43
44   [w, e] = eig(g);
45
46   [rg,~] = size(g);
47
48   es = sort(diag(e));
49
50   % Look for repeated roots
51   if ( length(unique(es)) < length(es) )
52     rr = 1;
53   else
54     rr = 0;
55   end
56
```

```
57   % Sort the columns of w
58   ws    = w(:,real(diag(e)) < 0);
59
60   sinf = real(ws(rg/2+1:rg,:)/ws(1:rg/2,:));
```

The demo is for a double integrator. This is just

$$\ddot{z} = u \tag{6.17}$$

In state space form, this becomes

$$\begin{bmatrix} \dot{z} \\ \dot{v} \end{bmatrix} = \begin{bmatrix} 0 & 1 \\ 0 & 0 \end{bmatrix} \begin{bmatrix} z \\ v \end{bmatrix} + \begin{bmatrix} 0 \\ 1 \end{bmatrix} u \tag{6.18}$$

The states are z, position, and v, velocity. u is the input. The following is the built-in demo code. The demo shows that the function creates a controller.

```
62   %% QCR>Demo
63   function Demo
64
65   a = [0 1;0 0];
66   b = [0;1];
67   q = eye(2);
68   r = 1;
69
70   k = QCR( a, b, q, r );
71
72   e = eig(a-b*k);
73
74   fprintf('\nGain = [%5.2f %5.2f]\n\n',k);
75   disp('Eigenvalues');
76   disp(e)
```

We chose the cost on the states and the control to all be 1.

```
>> QCR

Gain = [ 1.00  1.73]

Eigenvalues
  -0.8660 + 0.5000i
  -0.8660 - 0.5000i
```

We compute the eigenvalues to show that the result is well behaved. The result is critically damped, that is, damping ratio of 0.7071 in the second-order damped oscillator equation.

$$x^2 + 2\zeta\omega x + \omega^2 = 0 \tag{6.19}$$

6.3 Dynamical Model

6.3.1 Problem

Create a dynamical model of the Tokamak.

6.3.2 Solution

Implement the plasma dynamics model in a MATLAB function.

6.3.3 How It Works

We first code the right-hand-side function. We create the four matrices in the `DefaultData` `Structure` function. This makes the right-hand side simple.

RHSTokamak.m

```
 5   %% Inputs
 6   %   x     (3,1)  State [iA;iV;v] [active coils;passive coils;delay state]
 7   %   t     (1,1)  Time (s)
 8   %   d     (.)    Structure
 9   %                .aS   (2,2)  State matrix
10   %                .bS   (2,3)  Input matrix
11   %                .cS   (1,2)  Output matrix
12   %                .dS   (1,2)  Feed through matrix
13   %                .tauT (1,1)  Input time constant (s)
14   %                .vC   (1,1)  Control voltage
15   %                .eLM  (1,1)  Edge localized mode disturbance
16   %                .iP   (1,1)  Plasma current (A)
17   %
18   %% Outputs
19   %   xDot  (3,1)  State derivative d[iA;iV;v]/dt
20   %   z     (1,1)  Plasma position (m)
21   %
22   %% Reference:
23   % Scibile, L. "Non-linear control of the plasma vertical
24   % position in a tokamak," Ph.D Thesis, University of Oxford, 1997.
25
26   function [xDot,z] = RHSTokamak( x, ~, d )
27
28   if( nargin < 3 )
29     if( nargin == 1 )
30       xDot = UpdateDataStructure(x);
31     else
32       xDot = DefaultDataStructure;
33     end
34
35     return;
36   end
37
38   u    = [d.vC;d.eLM];
39   vDot = (x(3) - d.vC)/d.tauT;
```

```
40  xDot = [d.aS*x(1:2) + d.bS*u;vDot];
41  z    = d.cS*x(1:2) + d.dS*u;
42
43  function d = DefaultDataStructure
44
45  d = struct( 'lAA', 42.5e-3, 'lAV', 0.432e-3, 'lVV', 0.012e-3,...
46              'rAA', 35.0e-3, 'rVV',2.56e-3,'lAP',115.2e-6,'lVP',3.2e
                  -6,...
47              'aPP',0.449e-6,'tauT',310e-6,'iP',1.5e6,'aS',[],'bS',[],'cS
                  ',[],'dS',[],...
48              'eLM',0,'vC',0);
49
50  d = UpdateDataStructure( d );
51
52  function d = UpdateDataStructure( d )
53
54  kAV    = d.lAV^2/(d.lAA*d.lVV);
55  oMKAV  = 1 - kAV;
56  kA     = 1/(d.lAA*oMKAV);
57  mVP    = d.aPP*d.lVV/d.lVP^2;
58  oMMVP  = 1 - mVP;
59
60  if( mVP >= 1 )
61    fprintf('mVP = %f should be less than 1 for an elongated plasma in a
          resistive vacuum vessel. aPP is probably too large\n',mVP);
62  end
63
64  if( kAV >= 1 )
65    fprintf('kAV = %f should be less than 1 for an elongated plasma in a
          resistive vacuum vessel\n',kAV);
66  end
67
68  d.aS   = (1/oMKAV)*[ -d.rAA/d.lAA d.rVV*kAV/d.lAV;...
69                        d.rAA*kAV/d.lAV -(d.rVV/d.lVV)*(kAV - mVP)/
                          oMMVP];
70  d.bS   = [kA 0 0;kAV/(d.lAV*(1-kAV)) 1/(d.lVP*oMMVP) 0];
71  d.cS   = -[d.lAP d.lVP]/d.aPP/d.iP;
72  d.dS   = [0 0 1]/d.aPP/d.iP;
73  eAS    = eig(d.aS);
74
75  disp('Eigenvalues')
76  fprintf('\n Mode 1 %12.2f\n Mode 2 %12.2f\n',eAS);
```

If we type RHSTokamak at the command line, we get the default data structure.

```
>> RHSTokamak

ans =

  struct with fields:
```

```
  lAA: 0.0425
  lAV: 4.3200e-04
  lVV: 1.2000e-05
  rAA: 0.0350
  rVV: 0.0026
  lAP: 1.1520e-04
  lVP: 3.2000e-06
  aPP: 4.0000e-07
 tauT: 3.1000e-04
   aS: [2x2 double]
   bS: [2x3 double]
   cS: [-288 -8]
   dS: [0 2500000 0]
  eLM: 0
   vC: 0
```

This is the four matrices plus all the constants. The two inputs, control voltage d.vC, and ELM disturbances, d.eLM, are zero. If you have your own values of lAA, and so on, you can do

```
d = RHSTokamak;
d.lAA = 0.046;
d.tauT = 0.00035;
d = RHSTokamak(d)
```

and it will create the matrices. There are two warnings to prevent you from entering invalid parameters.

To see that the system really is unstable, type

```
>> RHSTokamak
Eigenvalues

  Mode 1        -2.67
  Mode 2       115.16
  Delay     -3225.81
```

These agree with the reference for JET. The third is the first-order lag. Note that

```
>> d.aPP

ans =

   4.4900e-07
```

This value was chosen so that the roots match the JET numbers.

6.4   Simulate the Plasma

6.4.1   Problem

We want to simulate the vertical position dynamics of the plasma with ELM disturbances.

6.4.2   Solution

Write a simulation script called `DisruptionSim.m`.

6.4.3   How It Works

The simulation script is an open-loop simulation of the plasma.

DisruptionSim.m

```
15   %% Constants
16   d            = RHSTokamak;
17   tau1ELM      = 6.0e-4;    % ELM time constant 1
18   tau2ELM      = 1.7e-4;    % ELM time constant 2
19   kELM         = 1.5e-6;    % ELM gain matches Figure 2.9 in Reference 2
20   tRepELM      = 48e-3;     % ELM repetition time (s)
21
22   %% The control sampling period and the simulation integration time step
23   dT           = 1e-4;
24
25   %% Number of sim steps
26   nSim         = 1200;
27
28   %% Plotting array
29   xPlot        = zeros(7,nSim);
30
31   %% Initial conditions
32   x            = [0;0;0]; % State is zero
33   t            = 0; % % Time
34   tRep         = 0.001; % Time for the 1st ELM
35   tELM         = inf; % Prevents an ELM at the start
36   zOld         = 0; % For the first difference rate equation
37
38   %% Run the simulation
39   for k = 1:nSim
40     d.v   = 0;
41     d.eLM = ELM( tau1ELM, tau2ELM, kELM, tELM );
42     tELM  = tELM + dT;
43
44     % Trigger another ELM
45     if( t > tRep + rand*tRepELM )
46           tELM   = 0;
47           tRep   = t;
48     end
49
```

```
50    x              = RungeKutta( @RHSTokamak, t, x, dT, d );
51    [~,z]          = RHSTokamak( x, t, d );
52    t              = t + dT;
53    zDot           = (z - zOld)/dT;
54    xPlot(:,k)     = [x;z;zDot;d.eLM];
55  end
56
57  %% Plot the results
58  tPlot = dT*(0:nSim-1)*1000;
59  yL    = {'I_A' 'I_V' 'v' 'z (m)' 'zDot (m/s)' 'ELM' 'ELMDot'};
60  k     = [1 2 4 5];
61  PlotSet( tPlot, xPlot(k,:), 'x label', 'Time (ms)', 'y label', yL(k), '
          figure title', 'Disruption Simulation' );
62  k     = [5, 6];
63  PlotSet( tPlot, xPlot(k,:), 'x label', 'Time (ms)', 'y label', yL(k), '
          figure title', 'ZDot and ELM' );
```

It prints out the eigenvalues for reference to make sure the dynamics work correctly.

```
>> DisruptionSim
Eigenvalues

Mode 1        -2.67
Mode 2       115.16
Delay      -3225.81
```

The value for the magnitude of the ELMs was found by running the simulation and looking at the magnitude of $\dot{z}$ and matching them to the results in the reference.

```
20    tRepELM     = 48e-3;    % ELM repetition time (s)
```

The value of the time derivative of the plasma vertical position z is just the first difference.

```
53    zDot        = (z - zOld)/dT;
```

The ELMs are triggered randomly inside of the simulation loop. tRep is the time of the last ELM. It adds a random amount of time to this number.

```
45    if( t > tRep + rand*tRepELM )
46          tELM    = 0;
47          tRep    = t;
48    end
```

The results are shown in Figure 6.6. The currents grow with time due to the positive eigenvalue. The only disturbance is the ELMs, but they are enough to cause the vertical position to grow.

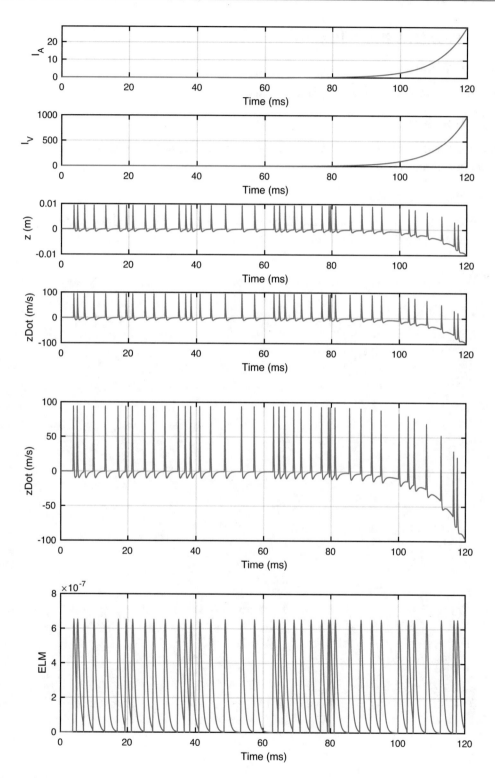

Figure 6.6: *Plasma simulation. The position is unstable.*

6.5 Control the Plasma

6.5.1 Problem

We want to control the plasma vertical position.

6.5.2 Solution

Write a simulation script called `ControlSim.m` to demonstrate closed loop control of the vertical position of the plasma.

6.5.3 How It Works

`ControlSim` is a closed-loop simulation of the plasma. We added the controller with the gains computed by the `QCR` function.

ControlSim.m

```
 1   %% Tokamak Plasma Vertical Control Demo
15   %% Constants
16   d            = RHSTokamak;
17   tau1ELM      = 6.0e-4;    % ELM time constant 1
18   tau2ELM      = 1.7e-4;    % ELM time constant 2
19   kELM         = 1.5e-6;    % ELM gain matches Figure 2.9 in Reference 2
20   tRepELM      = 48e-3;     % ELM repetition time (s)
21   controlOn    = true;
22   vCMax        = 3e-4;
23
24   %% The control sampling period and the simulation integration time step
25   dT           = 1e-5;
26
27   %% Number of sim steps
28   nSim         = 20000;
29
30   %% Plotting array
31   xPlot        = zeros(8,nSim);
32
33   %% Initial conditions
34   x            = [0;0;0];
35   t            = 0;
36   tRep         = 0.001; % Time for the 1st ELM
37   tELM         = inf; % This value will be change after the first ELM
38   zOld         = 0; % For the rate equation
39   z            = 0;
40
41   %% Design the controller
42   kControl     = QCR( d.aS, d.bS(:,1), eye(2), 1 );
43
```

```
44   %% Run the simulation
45   for k = 1:nSim
46     if( controlOn )
47       d.vC = -kControl*x(1:2);
48       if( abs(d.vC) > vCMax )
49         d.vC = sign(d.vC)*vCMax;
50       end
51     else
52       d.vC          = 0; %#ok<UNRCH>
53     end
54
55     d.eLM = ELM( tau1ELM, tau2ELM, kELM, tELM );
56     tELM  = tELM + dT;
57
58     % Trigger another ELM
59     if( t > tRep + rand*tRepELM )
60             tELM  = 0;
61             tRep  = t;
62     end
63
64     x             = RungeKutta( @RHSTokamak, t, x, dT,  d );
65     [~,z]         = RHSTokamak( x, t, d ); % Get the position
66     t             = t + dT;
67     zDot          = (z - zOld)/dT; % The rate of the vertical position
68     xPlot(:,k)    = [x;z;zDot;d.eLM;d.vC];
69   end
70
71   %% Plot the results
```

The controller is implemented in the loop. It applies the limiter.

```
46     if( controlOn )
47       d.vC = -kControl*x(1:2);
48       if( abs(d.vC) > vCMax )
49         d.vC = sign(d.vC)*vCMax;
50       end
51     else
52       d.vC          = 0; %#ok<UNRCH>
53     end
```

Results are shown in Figure 6.7.

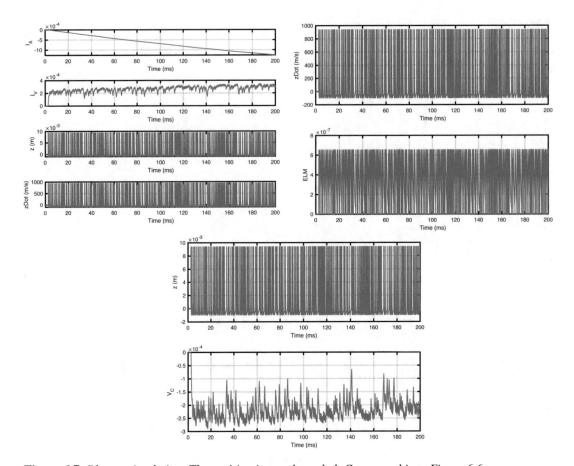

Figure 6.7: *Plasma simulation. The position is now bounded. Compare this to Figure 6.6.*

6.6   Training and Testing

6.6.1 Problem

We want to detect measurements leading up to disruptions.

6.6.2 Solution

We use a biLSTM (Bidirectional Long Short-Term Memory) layer to detect disruptions by classifying a time sequence as leading up to a disruption or not. LSTMs are designed to avoid dependency on old information. A standard RNN has a repeating structure. An LSTM also has a repeating structure, but each element has four layers. The LSTM layers decide what old information to pass on to the next layer. It may be all, or it may be none. There are many variants of LSTM, but they all include the fundamental ability to forget things. biLSTM is generally better than an LSTM when we have the full-time sequence.

6.6.3 How It Works

The following script, `TokamakNeuralNet.m`, generates the training and the test data, trains the neural network, and tests it. The constants are initialized first.

TokamakNeuralNet.m

```
1   %% Use LSTM to classify time sequences from the tokamak simulation
6   %% Constants
7   d          = RHSTokamak;
8   tau1ELM    = 6.0e-4;    % ELM time constant 1
9   tau2ELM    = 1.7e-4;    % ELM time constant 2
10  kELM       = 1.5e-6;    % ELM gain matches Figure 2.9 in Reference 2
11  tRepELM    = 48e-3;     % ELM repetition time (s)
12  controlOn  = true;      % Turns on the controller
13  disThresh  = 1.6e-6;    % This is the threshold for a disruption
14
15  % The control sampling period and the simulation integration time step
16  dT = 1e-5;
17
18  % Number of sim steps
19  nSim = 2000;
20
21  % Number of tests
22  n          = 100;
23  sigma1ELM = 2e-6*abs(rand(1,n));
24
25  PlotSet(1:n,sigma1ELM,'x label','Test Case','y label','1 \sigma ELM
        Value');
26
27  zData      = zeros(1,nSim); % Storage for vertical position
```

We design the controller as we did in `ControlSim`. The script runs 100 simulations. The linear quadratic controller demonstrated in `ControlSim` controls the position.

```
29  %% Initial conditions
30  x    = [0;0;0]; % The state of the plasma
31  tRep = 0.001; % Time for the 1st ELM
32
33  %% Design the controller
34  kControl = QCR( d.aS, d.bS(:,1), eye(2), 1 );
35
36  s          = cell(n,1);
37
38  %% Run n simulation
39  for j = 1:n
40          % Run the simulation
41      t    = 0;
```

```
42    tELM    = inf; % Prevents an ELM at the start
43         kELM   = sigma1ELM(j);
44    tRep    = 0.001; % Time for the 1st ELM
45
46    for k = 1:nSim
47      if( controlOn )
48        d.vC = -kControl*x(1:2);
49      else
50        d.vC        = 0; %#ok<UNRCH>
51      end
52
53      d.eLM        = ELM( tau1ELM, tau2ELM, kELM, tELM );
54      tELM         = tELM + dT;
55
56      % Trigger another ELM
57      if( t > tRep + rand*tRepELM )
58        tELM       = 0;
59        tRep       = t;
60      end
61
62      x            = RungeKutta( @RHSTokamak, t, x, dT, d );
63      [~,z]        = RHSTokamak( x, t, d );
64      t            = t + dT;
65      zData(1,k) = z;
66    end
67    s{j} = zData;
68  end
69
70  clear c
```

A disruption is any time response with z peaks over the threshold. Figure 6.8 shows the responses and the distribution of standard deviations. The blue line is a simulation that failed to keep vertical displacement of the plasma field under the prescribed threshold, and the red line is a simulation that succeeded in keeping the displacements below that threshold. The classification criteria are set in the following code.

```
72  %% Classify the results
73  j         = find(sigma1ELM > disThresh);
74  jN        = find(sigma1ELM < disThresh);
75  c(j,1)    = 1;
76  c(jN,1)   = 0;
77
78  [t,tL] = TimeLabel((0:nSim-1)*dT);
79  PlotSet(t,[s{j(1)};s{jN(1)}],'x label',tL,'y label','z (m)','Plot Set'
         ,{1:2},'legend',{{'disruption','stable'}});
```

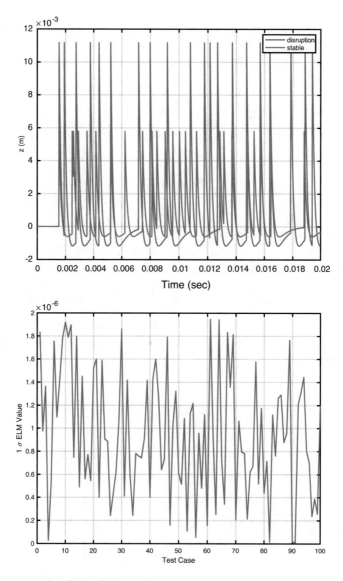

Figure 6.8: *Time responses and distribution of one-sigma values.*

The training is done next.

```
81   %% Divide into training and testing data
82   nTrain   = floor(0.8*n); % Train on 80% of the cases
83   xTrain   = s(1:nTrain);
84   yTrain   = categorical(c(1:nTrain));
85   xTest    = s(nTrain+1:n);
86   yTest    = categorical(c(nTrain+1:n));
87
88   %% Train the neural net
```

```
89   numFeatures     = 1; % Just the plasma position
90   numClasses      = 2; % Disruption or non disrupton
91   numHiddenUnits  = 200;
92
93   layers = [ ...
94       sequenceInputLayer(numFeatures)
95       bilstmLayer(numHiddenUnits,'OutputMode','last')
96       fullyConnectedLayer(numClasses)
97       softmaxLayer
98       classificationLayer];
99   disp(layers)
100
101  options = trainingOptions('adam', ...
102      'MaxEpochs',60, ...
103      'GradientThreshold',2, ...
104      'Verbose',0, ...
105      'Plots','training-progress');
106
107  net = trainNetwork(xTrain,yTrain,layers,options);
```

The training is shown in Figure 6.9. It takes over 42 minutes and converges after 10 epochs. Given this, if we rerun training with fewer MaxEpochs – perhaps 20 – training will be faster; this is left to the reader.

The testing is done next.

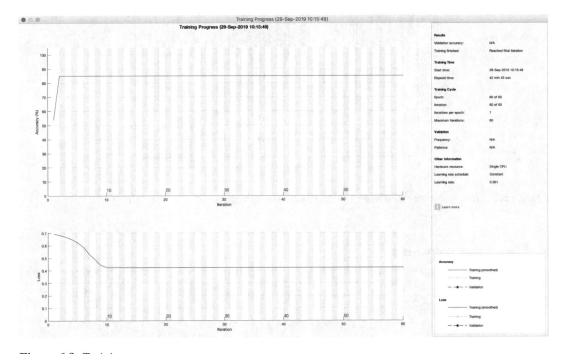

Figure 6.9: *Training.*

```
109  %% Demonstrate the neural net
110
111  %%% Test the network
112  yPred    = classify(net,xTest);
113
114  % Calculate the classification accuracy of the predictions.
115  acc           = sum(yPred == yTest)./numel(yTest);
116  disp('Accuracy')
117  disp(acc);
```

The results are encouraging. The International Thermonuclear Experimental Reactor (ITER) will require 95% of disruption predictions to be correct and to present an alarm 30 ms before a disruption [34]. Good results have been obtained using data from DIII-D [19].

```
>> TokamakNeuralNet
Eigenvalues

  Mode 1         -2.67
  Mode 2         115.16
   5x1 Layer array with layers:

       1    ''    Sequence Input          Sequence input with 1 dimensions
       2    ''    BiLSTM                  BiLSTM with 200 hidden units
       3    ''    Fully Connected         2 fully connected layer
       4    ''    Softmax                 softmax
       5    ''    Classification Output   crossentropyex
  Accuracy
       0.7500
```

This chapter did not deal with recursive or online training. A disruption prediction would need to constantly incorporate new data into its neural network. In addition, the other criteria for disruption detection would also need to be incorporated.

CHAPTER 7

■ ■ ■

Classifying a Pirouette

7.1   Introduction

A pirouette is a familiar step in ballet. There are many types of pirouettes. We will focus on an en dehors (outside) pirouette from the fourth position. The dancer pliés (does a deep knee bend) and then straightens their legs, producing both an upward force to get on the tip of their pointe shoe and torque to turn about their axis of revolution.

In this chapter, we will classify pirouettes. The deep learning neural network will classify pirouettes by dancer. Four dancers will each do ten double pirouettes, and we will use them to train the deep learning network. The network can then be used to classify pirouettes.

This chapter will involve real-time data acquisition and deep learning. We will spend a considerable amount of time in this chapter creating software to interface with the hardware. While it is not deep learning, it is important to know how to get data from sensors for use in deep learning work. We give code snippets in this chapter. Only a few can be cut and pasted into the MATLAB command window. You'll need to run the software in the downloadable library. Also remember, you will need the Instrument Control toolbox for this project.

Our subject dancers showing a pirouette are shown in Figure 7.1. We have three female dancers and one male dancer. Two of the women are wearing pointe shoes. The measurements will be accelerations, angular rates, and orientation. There isn't any limit to the movements the dancers could do. We asked them all to do double pirouettes starting from the fourth position and returning to the fourth position. The fourth position is with one foot behind the other and separated by a quarter meter or so. This is in contrast to the fifth position where the feet are right against each other. Each is shown at the beginning, middle, and end of the turn. All have slightly different positions, though all are doing very good pirouettes. There is no one "right" pirouette. If you were to watch the turns, you would not be able to see that they are that different. The goal is to develop a neural network that can classify pirouettes.

This kind of tool would be useful in any physical activity. An athlete could train a neural network to learn any important movement. For example, a baseball pitcher's pitch could be learned. The trained network could be used to compare the same movement at any other time

Figure 7.1: *Dancers doing pirouettes. The stages are from left to right.*

to see if it has changed. A more sophisticated version, possibly including vision, might suggest how to fix problems or identify what has changed. This would be particularly valuable for rehabilitation.

7.1.1 Inertial Measurement Unit

Our sensing means will be the LPMS-B2 IMU that has Bluetooth. The range is sufficient to work in a ballet studio.

The IMU has many other outputs that we will not use. A close-up of the IMU is shown in Figure 7.2. Table 7.1 gives the parameters for the IMU.

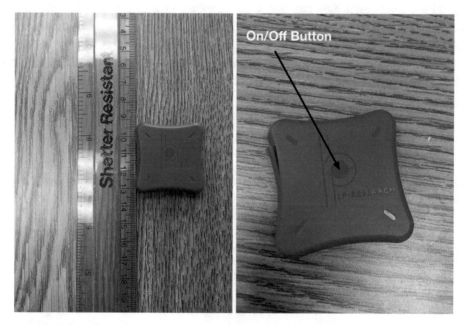

Figure 7.2: *LPMS-B2: 9-Axis Inertial Measurement Unit (IMU). The on/off button is highlighted on the right.*

Table 7.1: *LPMS-B2: 9-Axis Inertial Measurement Unit (IMU).*

Parameter	Description
Bluetooth	2.1 + EDR / Low Energy (LE) 4.1
Communication distance	< 20 m
Orientation range	Roll: ±180°; Pitch: ±90°; Yaw: ±180°
Resolution	< 0.01°
Accuracy	< 0.5°(static), < 2° RMS (dynamic)
Accelerometer	3 axes, ±2/ ± 4/ ± 8/ ± 16 g, 16 bits
Gyroscope 3 axes	±125/ ± 245/ ± 500/ ± 1000/ ± 2000°/s, 16 bits
Data output format	2 raw data / Euler angle / quaternion
Data transmission rate up to	2400Hz

We will first work out the details of the data acquisition. We will then build a deep learning algorithm to train the system and later take data and classify the pirouette as being a pirouette done by a particular dancer. We'll build up the data acquisition by first writing the MATLAB code to acquire the data. We will then create functions to display the data. We will then integrate it all into a GUI. Finally, we will create a deep learning classification system.

7.1.2 Physics

A pirouette is a complex multiflexible body problem. The pirouette is initiated by the dancer doing a plié and then using their muscles to generate a torque about the spin axis and forces to get onto their pointe shoe and over their center of mass. Their muscles quickly stop the translational motion so that they can focus on balancing as they are turning. We will assume that the dancer has a rigid body. Once a dancer is in the pirouette position, this is true. A dancer spots their head, that is, holds it focused on a target until the body has turned too far and then rapidly rotates it back to the target. However, we will ignore this for simplicity. The equation of rotational motion is known as Euler's equation and is

$$T = I\dot{\omega} + \omega^\times I\omega \tag{7.1}$$

where ω is the angular rate and I is the inertia. T is our external torque. The external torque is due to a push off the floor and gravity. This is a vector equation. The vectors are T and ω. I is a 3×3 matrix.

$$T = \begin{bmatrix} T_x \\ T_y \\ T_z \end{bmatrix} \tag{7.2}$$

$$\omega = \begin{bmatrix} \omega_x \\ \omega_y \\ \omega_z \end{bmatrix} \tag{7.3}$$

Each component is a value about a particular axis. For example, T_x is the torque about the x-axis attached to the dancer. Figure 7.3 shows the system. We will only be concerned with rotation assuming all translational motion is damped. If the dancer's center of mass is not above the box of their pointe shoe, they will experience an overturning torque.

The dynamical model is three first-order coupled differential equations. Angular rate ω is the state, that is, the quantity being differentiated. The equation says that the external torque (due to pushing off the floor or due to pointe shoe drag) is equal to the angular acceleration plus the Euler coupling term. This equation assumes that the body is rigid. For a dancer, it means they are rotating and no part is moving with respect to any other part. Now in a proper pirouette, this is never true if you are spotting. But let's suppose you are one of those dancers who don't spot. Let's forget about the angular rate coupling term, which only matters if the angular rate is

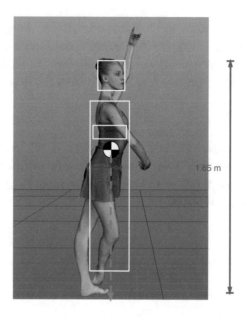

g

Figure 7.3: *The center of mass of the dancer. External forces act at the center of mass. Rotations are about the center of mass.*

large. Let's just look at the first two terms which is $T = I\dot{\omega}$. Expanded

$$
\begin{bmatrix} T_x \\ T_y \\ T_z \end{bmatrix} = \begin{bmatrix} I_{xx} & I_{xy} & I_{xz} \\ I_{xy} & I_{yy} & I_{yz} \\ I_{xz} & I_{yz} & I_{zz} \end{bmatrix} \begin{bmatrix} \dot{\omega}_x \\ \dot{\omega}_y \\ \dot{\omega}_z \end{bmatrix} \tag{7.4}
$$

Let's look at the equation for T_z. We just multiply the first row of the inertia matrix times the angular rate vector:

$$
T_z = I_{xz}\dot{\omega}_x + I_{yz}\dot{\omega}_y + I_{zz}\dot{\omega}_z \tag{7.5}
$$

This means a torque around the z-axis influences the angular rates about all three axes.

We can write out the required torque:

$$
\begin{bmatrix} T_x \\ T_y \\ T_z \end{bmatrix} = \begin{bmatrix} I_{xz} \\ I_{yz} \\ I_{zz} \end{bmatrix} \dot{\omega}_z \tag{7.6}
$$

This is the perfect pirouette push-off because it only creates rotation about the vertical axis, which is what we want in a pirouette. To this, you need to add the forces needed to get on pointe with your center of mass over your pointe shoe tip.

While turning, the only significant external torque is due to friction between the pointe shoe tip and the floor. Friction resists both turning motion and translational motion. You don't want a slight side force, perhaps due to a less than a great partner, to cause you to slide.

Our IMU measures angular rates and linear accelerations. Angular rates are the quantities in the Euler equations. However, since the IMU is not at the dancer's center of mass, it will also measure angular accelerations along with the acceleration of the center of mass. We locate it at the dancer's waist so it is not too far from the spin axis, but it still sees a component:

$$a_\omega = r_{\text{IMU}}^\times \dot{\omega} \tag{7.7}$$

where r_{IMU} is the vector from the dancer's center of mass to the IMU.

For a dancer doing a pirouette, Euler's equation is not sufficient. A dancer can transfer momentum internally to stop a pirouette and needs a little jump to get on demi-pointe or pointe. To model this, we add additional terms:

$$T = I\dot{\omega} + \omega^\times \left[I\omega + uI_i(\Omega_i + \omega_z) + uI_h(\Omega_h + \omega_z) \right] + u(T_i + T_h) \tag{7.8}$$

$$T_i = I_i \left(\dot{\Omega}_i + \dot{\omega}_z \right) \tag{7.9}$$

$$T_h = I_h \left(\dot{\Omega}_h + \dot{\omega}_z \right) \tag{7.10}$$

$$F = m\ddot{z} \tag{7.11}$$

where m is the mass, F is the vertical force, and z is the vertical direction. The vertical force is the difference between the dancer's leg force upward and gravity downward. I_i is the internal inertia for control, and I_s is the head inertia. I includes both of these already. That is, I is the total body inertia that includes the internal "wheel," body and head. T_i is the internal torque. T_h is the head torque (for spotting). The internal torques, T_i and T_h, are between the body and the internal "wheel" or head. For example, T_h causes the head to move one way and the body the other. If you are standing, the torque you produce from your feet against the floor prevents your body from rotating. T is the external torque due to friction and the initial push-off by the feet. The unit vector is

$$u = \begin{bmatrix} 0 \\ 0 \\ 1 \end{bmatrix} \tag{7.12}$$

There are six equations in total. The first is a vector equation with three components; the second two are scalar equations. The vector equation is three equations, and each scalar equation is just one equation. We can use these to create a simulation of a dancer. The second component models all z-axis internal rotation, including spotting.

7.2 Data Acquisition

7.2.1 Problem

We want to get data from the Bluetooth IMU.

7.2.2 Solution

We will use the MATLAB `bluetooth` function. We'll create a function to read data from the IMU.

7.2.3 How It Works

We will write an interface for the Bluetooth device. First, make sure the IMU is charged. Connect it to your computer as shown in Figure 7.4. Push the button on the back. This turns it on and off. The status is indicated by the LED. The IMU comes with support software from the vendor, but you will not need any of their software as MATLAB does all the hard work for you.

Let's try commanding the IMU. Type `btInfo = instrhwinfo('Bluetooth')` and you should get the following:

```
>> btInfo = instrhwinfo('Bluetooth')

btInfo =

  HardwareInfo with properties:

          RemoteNames: {'LPMSB2-4B31D6'}
            RemoteIDs: {'btspp://00043E4B31D6'}
      BluecoveVersion: 'BlueCove-2.1.1-SNAPSHOT'
       JarFileVersion: 'Version 4.0'
```

Figure 7.4: *The IMU is connected to the Bluetooth device from a MacBook Pro via USB C and a Mac dongle. This is only for charging purposes. Once it is charged, you can use it without the dongle.*

```
Access to your hardware may be provided by a support package. Go to the
    Support Package Installer to learn more.
```

This shows that your IMU is discoverable. There is no support package available from Math-Works. Now type b = Bluetooth(btInfo.RemoteIDs1,1) (this can be slow). The number is the channel. The Bluetooth function requires the Instrument Control Toolbox for MATLAB.

```
>> b = Bluetooth(btInfo.RemoteIDs{1},1)

    Bluetooth Object : Bluetooth-btspp://00043E4B31D6:1

    Communication Settings
        RemoteName:         LPMSB2-4B31D6
        RemoteID:           btspp://00043e4b31d6
        Channel:            1
        Terminator:         'LF'

    Communication State
        Status:             closed
        RecordStatus:       off

    Read/Write State
        TransferStatus:     idle
        BytesAvailable:     0
        ValuesReceived:     0
        ValuesSent:         0
```

Note that the Communication State Status shows closed. The device is opened by typing fopen(b). If you don't have this device, just type

```
>> btInfo = instrhwinfo('Bluetooth')
btInfo =
  HardwareInfo with properties:

        RemoteNames: []
          RemoteIDs: []
    BluecoveVersion: 'BlueCove-2.1.1-SNAPSHOT'
     JarFileVersion: 'Version 4.0'
Access to your hardware may be provided by a support package. Go to the
    Support Package Installer to learn more.
```

This says it cannot recognize remote names or IDs. You may need a support package for your device in this case.

Click connect and the device will open. Now type a= fscanf(b) and you will get a bunch of unprintable characters. We now have to write code to command the device. We will leave the device in streaming mode. The data unit format is shown in Table 7.2. Each packet is 91 bytes long even though the table only shows 67 bytes. The 67 bytes are all the useful data.

Table 7.2: *Reply data.*

Byte	Content (hex)	Meaning
0	3A	Packet Start
1	01	OpenMAT ID LSB (ID=1)
2	00	OpenMAT MSB
3	09	Command No. LSB (9d = GET_SENSOR_DATA)
4	00	Command No. MSB
5	00	Data Length LSB
6	00	Data Length MSB
7–10	xxxxxxxx	Timestamp
11–14	xxxxxxxx	Gyroscope data x-axis
15–18	xxxxxxxx	Gyroscope data y-axis
19–22	xxxxxxxx	Gyroscope data z-axis
23–26	xxxxxxxx	Accelerometer x-axis
27–30	xxxxxxxx	Accelerometer y-axis
31–34	xxxxxxxx	Accelerometer z-axis
35–38	xxxxxxxx	Magnetometer x-axis
39–42	xxxxxxxx	Magnetometer y-axis
43–46	xxxxxxxx	Magnetometer z-axis
47–50	xxxxxxxx	Orientation quaternion q0
51–54	xxxxxxxx	Orientation quaternion q1
55–58	xxxxxxxx	Orientation quaternion q2
59–62	xxxxxxxx	Orientation quaternion q3
63	xx	Check sum LSB
64	xx	Check sum MSB
65	0D	Message end byte 1
66	0A	Message end byte 2

We read the binary and put it into a data structure using `DataFromIMU`. `typecast` converts from bytes to float.

DataFromIMU.m

```
25  function d = DataFromIMU( a )
26
27  d.packetStart    = dec2hex(a(1));
28  d.openMATIDLSB   = dec2hex(a(2));
29  d.openMATIDMSB   = dec2hex(a(3));
30  d.cmdNoLSB       = dec2hex(a(4));
31  d.cmdNoMSB       = dec2hex(a(5));
32  d.dataLenLSB     = dec2hex(a(6));
33  d.dataLenMSB     = dec2hex(a(7));
34  d.timeStamp      = BytesToFloat( a(8:11) );
```

```
35   d.gyro           = [ BytesToFloat( a(12:15) );...
36                        BytesToFloat( a(16:19) );...
37                        BytesToFloat( a(20:23) )];
38   d.accel          = [ BytesToFloat( a(24:27) );...
39                        BytesToFloat( a(28:31) );...
40                        BytesToFloat( a(32:35) )];
41   d.quat           = [ BytesToFloat( a(48:51) );...
42                        BytesToFloat( a(52:55) );...
43                        BytesToFloat( a(56:59) );...
44                        BytesToFloat( a(60:63) )];
45   d.msgEnd1        = dec2hex(a(66));
46   d.msgEnd2        = dec2hex(a(67));
48
49   %% DataFromIMU>BytesToFloat
50   function r = BytesToFloat( x )
51
52   r = typecast(uint8(x),'single');
```

We've wrapped all of this into the script BluetoothTest.m. We print out a few samples of the data to make sure our bytes are aligned correctly.

BluetoothTest.m

```
1    %% Script to read binary from the IMU
2
3    % Find available Bluetooth devices
4    btInfo = instrhwinfo('Bluetooth')
5
6     % Display the information about the first device discovered
7    btInfo.RemoteNames(1)
8    btInfo.RemoteIDs(1)
9
10   % Construct a Bluetooth Channel object to the first Bluetooth device
11   b = Bluetooth(btInfo.RemoteIDs{1}, 1);
12
13   % Connect the Bluetooth Channel object to the specified remote device
14   fopen(b);
15
16   % Get a data structure
17   tic
18   t = 0;
19   for k = 1:100
20     a   = fread(b,91);
21     d   = DataFromIMU( a );
22     fprintf('%12.2f [%8.1e %8.1e %8.1e] [%8.1e %8.1e %8.1e] [%8.1f %8.1f
         %8.1f %8.1f]\n',t,d.gyro,d.accel,d.quat);
23     t = t + toc;
24     tic
25   end
```

When we run the script we get the following output.

```
>> BluetoothTest

btInfo =

  HardwareInfo with properties:

        RemoteNames: {'LPMSB2-4B31D6'}
          RemoteIDs: {'btspp://00043E4B31D6'}
    BluecoveVersion: 'BlueCove-2.1.1-SNAPSHOT'
     JarFileVersion: 'Version 4.0'

Access to your hardware may be provided by a support package. Go to the
    Support Package Installer to learn more.

ans =

  1x1 cell array

    {'LPMSB2-4B31D6'}

ans =

  1x1 cell array

    {'btspp://00043E4B31D6'}

ans =

  1x11 single row vector

    1.0000    0.0014    0.0023   -0.0022    0.0019   -0.0105   -0.9896
          0.9200   -0.0037    0.0144    0.3915

ans =

  1x11 single row vector

    2.0000   -0.0008    0.0023   -0.0016    0.0029   -0.0115   -0.9897
          0.9200   -0.0037    0.0144    0.3915

ans =

  1x11 single row vector

    3.0000    0.0004    0.0023   -0.0025    0.0028   -0.0125   -0.9900
          0.9200   -0.0037    0.0144    0.3915
```

The first number in each row vector is the sample, the next three are the angular rates from the gyro, the next three are the accelerations, and the last four are the quaternion. The acceleration is mostly in the $-z$ direction which means that $+z$ is in the button direction. Bluetooth, like all wireless connections, can be problematic. If you get this error

```
Index exceeds the number of array elements (0).

Error in BluetoothTest (line 7)
btInfo.RemoteNames(1)
```

turn the IMU on and off. You might also have to restart MATLAB at times. This is because RemoteNames is empty, and this test is assuming it will not be. MATLAB then gets confused.

7.3 Orientation

7.3.1 Problem

We want to use quaternions to represent the orientation of our dancers in our deep learning system.

7.3.2 Solution

Implement basic quaternion operations. We need quaternion operations to process the quaternions from the IMU.

7.3.3 How It Works

Quaternions are the preferred mathematical representation of orientation. Propagating a quaternion requires fewer operations than propagating a transformation matrix and avoids singularities that occur with Euler angles. A quaternion has four elements, which corresponds to a unit vector a and angle of rotation ϕ about that vector. The first element is termed the "scalar component" s, and the next three elements are the "vector" components v. This notation is shown as follows [29]:

$$
q = \begin{bmatrix} q_0 \\ q_1 \\ q_2 \\ q_3 \end{bmatrix} = \begin{bmatrix} s \\ v_1 \\ v_2 \\ v_3 \end{bmatrix} = \begin{bmatrix} \cos \frac{\phi}{2} \\ a_1 \sin \frac{\phi}{2} \\ a_2 \sin \frac{\phi}{2} \\ a_3 \sin \frac{\phi}{2} \end{bmatrix} \tag{7.13}
$$

The "unit" quaternion which represents zero rotation from the initial coordinate frame has a unit scalar component and zero vector components. This is the same convention used on the Space Shuttle, although other conventions are possible.

$$
q_0 = \begin{bmatrix} 1 \\ 0 \\ 0 \\ 0 \end{bmatrix} \tag{7.14}
$$

In order to transform a vector from one coordinate frame a to another b using a quaternion q_{ab}, the operation is

$$u_b = q_{ab}^T u_a q_{ab} \tag{7.15}$$

using quaternion multiplication with the vectors defined as quaternions with a scalar part equal to zero, or

$$x_a = \begin{bmatrix} 0 \\ x_a(1) \\ x_a(2) \\ x_a(3) \end{bmatrix} \tag{7.16}$$

For example, the quaternion

$$\begin{bmatrix} 0.7071 \\ 0.7071 \\ 0.0 \\ 0.0 \end{bmatrix} \tag{7.17}$$

represents a pure rotation about the x-axis. The first element is 0.7071 and equals the $\cos(90°/2)$. We cannot tell the direction of rotation from the first element. The second element is the first component of the unit vector, which in this case is

$$\begin{bmatrix} 1.0 \\ 0.0 \\ 0.0 \end{bmatrix} \tag{7.18}$$

times the argument $\sin(90°/2)$. Since the sign is positive, the rotation must be a positive $90°$ rotation.

We only need one routine that converts the quaternion, which comes from the IMU, into a transformation matrix for visualization. We do this because multiplying a $3\times$ n array of vectors for the vertices of our 3D model by a matrix is much faster than transforming each vector with a quaternion.

QuaternionToMatrix.m

```
1   %% QUATERNIONTOMATRIX Converts a quaternion to a transformation matrix.
16  function m = QuaternionToMatrix( q )
17
18  m = [ q(1)^2+q(2)^2-q(3)^2-q(4)^2,...
19          2*(q(2)*q(3)-q(1)*q(4)),...
20          2*(q(2)*q(4)+q(1)*q(3));...
21          2*(q(2)*q(3)+q(1)*q(4)),...
22          q(1)^2-q(2)^2+q(3)^2-q(4)^2,...
23          2*(q(3)*q(4)-q(1)*q(2));...
24          2*(q(2)*q(4)-q(1)*q(3)),...
25          2*(q(3)*q(4)+q(1)*q(2)),...
26          q(1)^2-q(2)^2-q(3)^2+q(4)^2];
```

Note that the diagonal terms have the same form. The off-diagonal terms also all have the same form.

7.4 Dancer Simulation

7.4.1 Problem

We want to simulate a dancer for readers who don't have access to the hardware.

7.4.2 Solution

We will write a right-hand side for the dancer based on the preceding equations and write a simulation with a control system.

7.4.3 How It Works

The right-hand side implements the dancer model. It includes an internal control "wheel" and a degree of freedom for the head movement. The default data structure is returned if you call it without arguments.

RHSDancer.m

```
33  %% RHSDANCER Implements dancer dynamics
34  % This is a model of dancer with one degree of translational freedom
35  % and 5 degrees of rotational freedom including the head and an
        internal
36  % rotational degree of freedom.
37  %% Form:
38  %    xDot = RHSDancer( x, ~, d )
39  %% Inputs
40  %    x        (11,1)      State vector [r;v;q;w;wHDot;wIDot]
41  %    t        (1,1)    Time (unused) (s)
42  %    d        (1,1)    Data structure for the simulation
43  %                      .torque   (3,1) External torque (Nm)
44  %                      .force    (1,1) External force (N)
45  %                      .inertia  (3,3) Body inertia (kg-m^2)
46  %                      .inertiaH (1,1) Head inertia (kg-m^2)
47  %                      .inertiaI (1,1) Inner inertia (kg-m^2)
48  %                      .mass     (1,1) Dancer mass (kg)
49  %
50  %% Outputs
51  %    xDot     (11,1)      d[r;v;q;w;wHDot;wIDot]/dt
52
53  function xDot = RHSDancer( ~, x, d )
54
55  % Default data structure
56  if( nargin < 1 )
57     % Based on a 0.15 m radius, 1.4 m long cylinders
58     inertia = diag([8.4479     8.4479     0.5625]);
```

```
59    xDot     = struct('torque',[0;0;0],'force',0,'inertia',inertia,...
60    'mass',50,'inertiaI',0.0033,'inertiaH',0.0292,'torqueH',0,'torqueI'
          ,0);
61    return
62  end
```

The remainder mechanizes the equations given earlier. We add an equation for the integral of the z-axis rate. This makes the control system easier to write. We also include the gravitational acceleration in the force equation.

```
32  % Use local variables
33  v       = x(2);
34  q       = x(3:6);
35  w       = x(7:9);
36  wI      = x(10);
37  wH      = x(11);
38
39  % Unit vector
40  u       = [0;0;1];
41
42  % Gravity
43  g       = 9.806;
44
45  % Attitude kinematics (not mentioned in the text)
46  qDot    = QIToBDot( q, w );
47
48  % Rotational dynamics Equation 7.6
49  wDot    = d.inertia\(d.torque - Skew(w)*(d.inertia*w + d.inertiaI*(wI +
          w(3))...
50          + d.inertiaH*(wH + w(3))) - u*(d.torqueI + d.torqueH));
51  wHDot = d.torqueH/d.inertiaH - wDot(3);
52  wIDot = d.torqueI/d.inertiaI - wDot(3);
53
54  % Translational dynamics
55  vDot    = d.force/d.mass - g;
56
57  % Assemble the state vector
58  xDot    = [v; vDot; qDot; wDot; wHDot; wIDot; w(3)];
```

The simulation setup gets default parameters from RHSDancer.

```
4   d       = RHSDancer;
5   n       = 800;
6   dT      = 0.01;
7   xP      = zeros(16,n);
8   x       = zeros(12,1);
9   x(3)    = 1;
10  g       = 9.806;
11  dancer  = 'Robot_1';
```

It then sets up the control system. We use a proportional derivative controller for the z position and a rate damper to stop the pirouette. The position control is done by the foot muscles. The rate damping is our internal damper wheel.

```
13   % Control system for 2 pirouettes in 6 seconds
14   tPirouette   = 6;
15   zPointe      = 6*0.0254;
16   tPointe      = 0.1;
17   kP           = tPointe/dT;
18   omega        = 4*pi/tPirouette;
19   torquePulse  = d.inertia(3,3)*omega/tPointe;
20   tFriction    = 0.1;
21   a            = 2*zPointe/tPointe^2 + g;
22   kForce       = 1000;
23   tau          = 0.5;
24   thetaStop    = 4*pi - pi/4;
25   kTorque      = 200;
26   state        = zeros(10,n);
```

The simulation loop calls the right-hand side and the control system. We call RHS Dancer.m to get the linear acceleration.

```
29   %% Simulate
30   for k = 1:n
31       d.torqueH = 0;
32       d.torqueI = 0;
33
34       % Get the data for use in the neural netwoork
35       xDot = RHSDancer(0,x,d);
36
37       state(:,k) = [x(7:9);0;0;xDot(2);x(3:6)];
38
39       % Control
40       if( k < kP )
41           d.force  = d.mass*a;
42           d.torque = [0;0;torquePulse];
43       else
44           d.force  = kForce*(zPointe-x(1) -x(2)/tau)+ d.mass*g;
45           d.torque = [0;0;-tFriction];
46       end
47
48       if( x(12) > thetaStop )
49           d.torqueI = kTorque*x(9);
50       end
51
52       xP(:,k)   = [x;d.force;d.torque(3);d.torqueH;d.torqueI];
53       x         = RungeKutta(@RHSDancer,0,x,dT,d);
54   end
```

The control system includes a torque and force pulse to get the pirouette going.

```
39    % Control
40    if( k < kP )
41        d.force  = d.mass*a;
42        d.torque = [0;0;torquePulse];
43    else
44        d.force  = kForce*(zPointe-x(1) -x(2)/tau)+ d.mass*g;
45        d.torque = [0;0;-tFriction];
46    end
```

The remainder of the script plots the results and outputs the data, which would have come from the IMU, into a file.

Simulation results for a double pirouette are shown in Figure 7.5. We stop the turn at 6.5 seconds, hence the pulse.

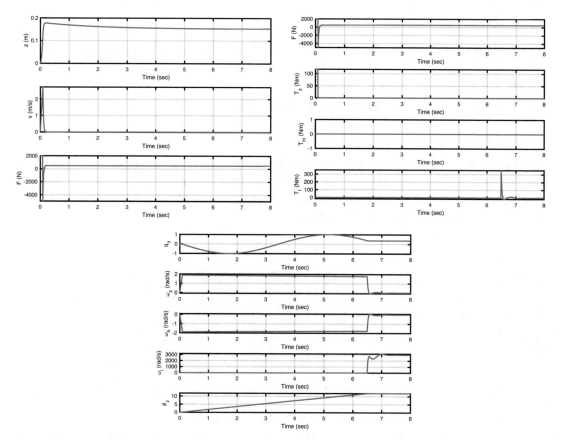

Figure 7.5: *Simulation of a double pirouette.*

You can create different dancers by varying the mass properties and the control parameters.

```
zPointe      = 6*0.0254;
tPointe      = 0.1;
tFriction    = 0.1;
kForce       = 1000;
tau          = 0.5;
kTorque      = 200;
```

We didn't implement spotting (looking at the audience as much as possible during the turn) control. It would rotate the head so that it faces forward whenever the head was within 90 degrees or so of the front. We'd need to add a head angle for that purpose to the right-hand side, much like we added the z-axis angle.

7.5   Real-Time Plotting

7.5.1 Problem

We want to display data from the IMU in real time. This will allow us to monitor the pirouettes.

7.5.2 Solution

Use `plot` with `drawnow` to implement multiple figures of plots.

7.5.3 How It Works

The main function is a switch statement with two cases. The function also has a built-in demo. The first case, `initialize`, initializes the plot figures. It stores everything in a data structure that is returned on each function call. This is one way for a function to have a memory. We return the data structure from each subfunction.

GUIPlots.m

```
31  switch( lower(action) )
32    case 'initialize'
33      g = Initialize( g );
34
35    case 'update'
36      g = Update( g, y, t );
37
38  end
```

The first case, `initialize`, initializes the figure window.

```
40  %% GUIPlots>Initialize
41  function g = Initialize( g )
42
43  lY = length(g.yLabel);
44
45  % Create tLim if it does not exist
```

```matlab
46  if( ~isfield(g, 'tLim' ) )
47      g.tLim = [0 1];
48  end
49
50  g.tWidth = g.tLim(2) - g.tLim(1);
51
52  % Create yLim if it does not exist
53  if( ~isfield( g, 'yLim' ) )
54      g.yLim = [-ones(lY,1), ones(lY,1)];
55  end
56
57  % Create the plots
58  lP = length(g.yLabel);
59  y  = g.pos(2); % The starting y position
60  for k = 1:lP
61      g.h(k) = subplot(lP,1,k);
62      set(g.h(k),'position',[g.pos(1) y g.pos(3) g.pos(4)]);
63      y = y - 1.4*g.pos(4);
64      g.hPlot(k) = plot(0,0);
65      g.hAxes(k) = gca;
66      g.yWidth(k) = (g.yLim(k,2) - g.yLim(k,1))/2;
67      set(g.hAxes(k),'nextplot','add','xlim',g.tLim);
68      ylabel( char(g.yLabel{k}) )
69      grid on
70  end
71  xlabel( g.tLabel );
```

The second case, update, updates the data displayed in the plot. It leaves the existing figures, subplots, and labels in place and just updates the plots of the line segments with new data. It can change the size of the axes as needed. The function adds a line segment for each new data point. This way, no storage is needed external to the plot. It reads xdata and ydata and appends the new data to those arrays.

```matlab
75  function g = Update( g, y, t )
76
77  % See if the time limits have been exceeded
78  if( t > g.tLim(2) )
79      g.tLim(2)   = g.tWidth + g.tLim(2);
80      updateAxes = true;
81  else
82      updateAxes = false;
83  end
84
85  lP = length(g.yLabel);
86  for k = 1:lP
87      subplot(g.h(k));
88      yD = get(g.hPlot(k),'ydata');
89      xD = get(g.hPlot(k),'xdata');
90      if( updateAxes )
91          set( gca, 'xLim', g.tLim );
```

```
92      set( g.hPlot(k), 'xdata',[xD t],'ydata',[yD y(k)]);
93    else
94      set( g.hPlot(k), 'xdata',[xD t],'ydata',[yD y(k)] );
95    end
96
97  end
```

The built-in demo plots six numbers. It updates the axes in time once. It sets up a figure window with six plots. You need to create the figure and save the figure handle before calling GUIPlots.

```
  g.hFig  = NewFig('State');
```

The pause in the demo just slows down plotting so that you can see the updates. The height (the last number in g.pos) is the height of each plot. If you happen to set the locations of the plots out of the figure window, you will get a MATLAB error. g.tLim gives the initial time limits in seconds. The upper limit will expand as data is entered.

```
100  function Demo
101
102  g.yLabel = {'x' 'y' 'z' 'x_1' 'y_1' 'z_1'};
103  g.tLabel = 'Time (sec)';
104  g.tLim   = [0 100];
105  g.pos    = [0.100    0.88    0.8    0.10];
106  g.width  = 1;
107  g.color  = 'b';
108
109  g.hFig   = NewFig('State');
110  set(g.hFig, 'NumberTitle','off' );
111
112  g          = GUIPlots( 'initialize', [], [], g );
113
114  for k = 1:200
115      y = 0.1*[cos((k/100))-0.05;sin(k/100)];
116      g = GUIPlots( 'update', [y;y.^2;2*y], k, g );
117      pause(0.1)
118  end
119
120  g          = GUIPlots( 'initialize', [], [], g );
121
122  for k = 1:200
123      y = 0.1*[cos((k/100))-0.05;sin(k/100)];
124      g = GUIPlots( 'update', [y;y.^2;2*y], k, g );
125      pause(0.1)
126  end
```

Figure 7.6 shows the real-time plots at the end of the demo.

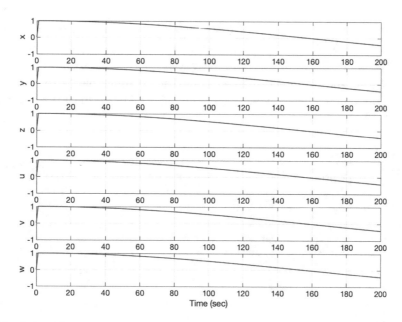

Figure 7.6: *Real-time plots.*

7.6   Quaternion Display

7.6.1 Problem

We want to display the dancer's orientation in real time.

7.6.2 Solution

Use `patch` to draw an OBJ model in a three-dimensional plot. The figure is easier to understand than the four quaternion elements. Our solution can handle three-axis rotation although typically we will only see single-axis rotation.

7.6.3 How It Works

We start with our `Ballerina.obj` file. It only has vertices and faces. A 3D drawing consists of a set of vertices. Each vertex is a point in space. The vertices are organized into faces. Each face is a triangle. Triangles are used for 3D drawings because they always form a plane. 3D processing hardware is designed to work with triangles, so this also gives the fastest results. The obj files for our software can only contain triangles. Each face can have only three vertices. Generally, obj files can have any size polynomials, that is, faces with more than three points. Most sources of obj files can provide tessellation services to convert polygons with more than three vertices into triangles. `LoadOBJ.m` will not draw models with anything other than triangular faces.

The main part of the function uses a case statement to handle the three actions. The first action just returns the defaults, which is the name of the default obj file. The second reads in the file and initializes the patches. The third updates the patches. `patch` is the MATLAB name for a set of triangles. The function can be passed through a figure handle. A figure handle tells it into which figure the 3D model should be drawn. This allows it to be used as part of a GUI, as well shown in the next section.

QuaternionVisualization.m

```
19  function m = QuaternionVisualization( action, x, f )
20
21  persistent p
22
23  % Demo
24  if( nargin < 1 )
25      Demo
26      return
27  end
28
29  switch( lower(action) )
30      case 'defaults'
31          m = Defaults;
32
33      case 'initialize'
34          if( nargin < 2 )
35              d   = Defaults;
36          else
37              d   = x;
38          end
39
40          if( nargin < 3 )
41              f = [];
42          end
43
44          p = Initialize( d, f );
45
46      case 'update'
47          if( nargout == 1 )
48              m = Update( p, x );
49          else
50              Update( p, x );
51          end
52  end
```

Initialize loads the obj file. It creates a figure and saves the object data structure. It sets shading to interpolated and lighting to Gouraud. Gouraud is a type of lighting model named after its inventor. It then creates the patches and sets up the axis system. We save handles to all the patches for updating later. We also place a light near the object to illuminate the patches.

```
55   function p = Initialize( file, f )
56
57   if( isempty(f) )
58     p.fig = NewFigure( 'Quaternion' );
59   else
60     p.fig = f;
61   end
62
63   g      = LoadOBJ( file );
64   p.g    = g;
65
66   shading interp
67   lighting gouraud
68
69   c = [0.3 0.3 0.3];
70
71   for k = 1:length(g.component)
72     p.model(k)    = patch('vertices', g.component(k).v, 'faces', g.
           component(k).f, 'facecolor',c,'edgecolor',c,'ambient',1,'
           edgealpha',0 );
73   end
74
75   xlabel('x');
76   ylabel('y');
77   zlabel('z');
78
79   grid
80   rotate3d on
81   set(gca,'DataAspectRatio',[1 1 1],'DataAspectRatioMode','manual')
82
83   light('position',10*[1 1 1])
84
85   view([1 1 1])
```

In Update, we convert the quaternion to a matrix because it is faster to matrix multiply all the vertices with one matrix multiplication. The vertices are n by 3 so we transpose before the matrix multiplication. We use the patch handles to update the vertices. The two options at the end are to create movie frames or just update the drawing.

```
88   function m = Update( p, q )
89
90   s = QuaternionToMatrix( q );
91
92   for k = 1:length(p.model)
93     v = (s*p.g.component(k).v')';
94     set(p.model(k),'vertices',v);
95   end
96
97   if( nargout > 0 )
98           m = getframe;
99   else
100          drawnow;
101  end
```

This is the built-in demo. We vary the 1 and 4 elements of the quaternion to get rotation about the z-axis.

```
109  function Demo
110
111  QuaternionVisualization( 'initialize', 'Ballerina.obj' );
```

Figure 7.7 shows two orientations of the dancer during the demo. The demo produces an animation of the dancer rotating about the z-axis. The rotation is slow because of the number of vertices. The figure is not articulated so the entire figure is rotated as a rigid body. We don't employ texture maps. In any case, the purpose of this function is just to show orientation so it doesn't matter.

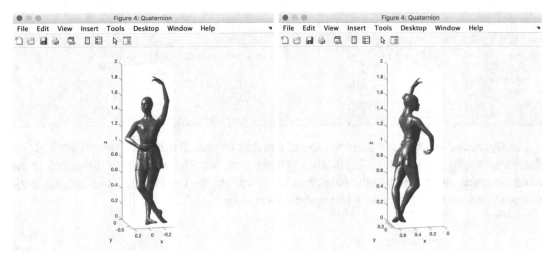

Figure 7.7: *Dancer orientation. The obj file is by the artists loft_22 and is available from TurboSquid.*

7.7 Making the IMU Belt

7.7.1 Problem

We need to attach the IMU to our dancer.

7.7.2 Solution

We use the arm strap that is available from the manufacturer. We buy an elastic belt and make one that fits around the dancer's waist.

7.7.3 How It Works

Yes, software engineers need to sew. Figure 7.8 shows the process. The two products used to make the data acquisition belt are

1. LPMS-B2 Holder (available from Life Performance Research)

2. Men's No Show Elastic Stretch Belt Invisible Casual Web Belt Quick Release Flat Plastic Buckle (available from Amazon)

Remove the holder from the LPMS-B2 Holder. Cut the belt at the buckle and slide the holder onto the belt. Sew the belt at the buckle.

The sensor on a dancer is shown in Figure 7.9. We had the dancer stand near the laptop during startup. We didn't have any range problems during the experiments. We didn't try it with across-the-floor movement as one would have during grande allegro.

Figure 7.8: *Elastic belt manufacturing. We use the left two items to make the one on the right.*

Figure 7.9: *Dancer with the sensor belt. The blue light means it is collecting data.*

7.8   Testing the System

7.8.1 Problem

We want to test the data acquisition system. This will find any problems with the data acquisition process.

7.8.2 Solution

Have a dancer do changements, which are small jumps changing the foot position on landing.

7.8.3 How It Works

The dancer puts on the sensor belt, we push the calibrate button, then they do a series of changements. The dancer stands about 2 m from your computer to make acquisition easier. The dancer will do small jumps, known as changements. A changement is a small jump where the feet change positions starting from the fifth position. If the right foot is in the fifth position front at the start, it is in the back at the finish. Photos are shown in Figure 7.10.

The time scale is a bit long. You can see that the calibration does not lead us to a natural orientation in the axis system in the GUI. It doesn't matter from a data collection point of view but is an improvement we should make in the future. The changement is shown in Figure 7.11. The dancer is still at the beginning and end.

Figure 7.10: *Dancer doing a changement. Notice the feet when the dancer is preparing to jump. The second image shows her feet halfway through the jump.*

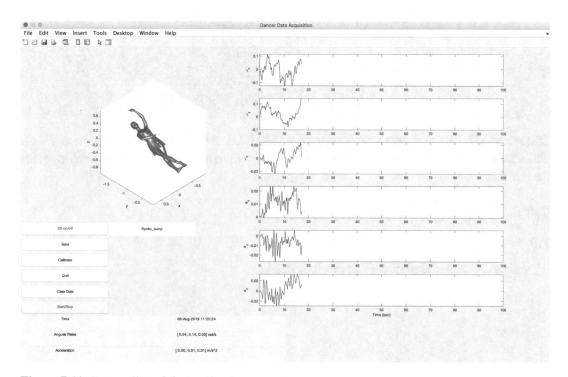

Figure 7.11: *Data collected during the changement.*

The interface to the Bluetooth device doesn't do any checking or stream control. Some Bluetooth data collection errors occur from time to time. Typically, they happen after 40 seconds of data collection.

```
index exceeds the number of array elements (0).
Error in instrhwinfo>bluetoothCombinedDevices (line 976)
        uniqueBTName = allBTName(uniqueRowOrder);
Error in instrhwinfo (line 206)
                tempOut = bluetoothCombinedDevices(tempOut);
Error in DataAcquisition (line 13)
btInfo  = instrhwinfo('Bluetooth');
```

If this happens, turn the device on and off. Restart MATLAB if that doesn't work. Another bluetooth error is

```
ans =
  1x1 cell array
    {'LPMSB2-4B31D6'}
ans =
  1x1 cell array
    {'btspp://00043E4B31D6'}
Error using Bluetooth (line 104)
Cannot Create: Java exception occurred:
java.lang.NullPointerException
        at com.mathworks.toolbox.instrument.BluetoothDiscovery.
          searchDevice(BluetoothDiscovery.java:395)
        at com.mathworks.toolbox.instrument.BluetoothDiscovery.
          discoverServices(BluetoothDiscovery.java:425)
        at com.mathworks.toolbox.instrument.BluetoothDiscovery.
          hardwareInfo(BluetoothDiscovery.java:343)
        at com.mathworks.toolbox.instrument.Bluetooth.<init>(Bluetooth.
          java:205).
```

This is a MATLAB error and requires restarting MATLAB. It doesn't happen very often. We ran the entire data collection with four dancers doing ten pirouettes each without ever experiencing the problem.

7.9   Classifying the Pirouette

7.9.1 Problem

We want to classify the pirouettes of our four dancers.

140

7.9.2 Solution

Create an LSTM that classifies pirouettes according to the dancer. The four labels are the dancers' names.

7.9.3 How It Works

The script takes one file and displays it. Figure 7.12 shows a double pirouette. It lasts only a few seconds.

DancerNN.m

```
1  %% Script to train and test the dancer neural net
5  dancer = {'Ryoko' 'Shaye', 'Emily', 'Matanya'};
6
7  %% Show one dancer's data
8  cd TestData
9  s    = load('Ryoko_10.mat');
10 yL   = {'\omega_x' '\omega_y' '\omega_z' 'a_x' 'a_y' 'a_z'};
11 PlotSet(s.time,s.state(1:6,:),'x label','Time (s)','y label',yL,'figure
         title',dancer{1});
```

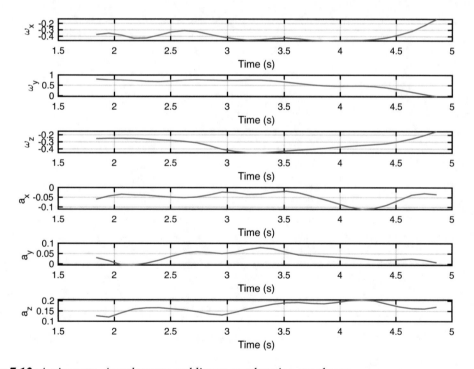

Figure 7.12: *A pirouette. Angular rate and linear acceleration are shown.*

We load in the data and limit the range to six seconds. Sometimes, the IMU would run longer due to human error (the data collection did not stop after the pirouette ended). We also remove bad sets.

DancerNN.m

```
14   %% Load in and process the data
15   n = 40;
16   % Get the data and remove bad data sets
17   i = 0;
18   for k = 1:length(dancer)
19     for j = 1:10
20       s   = load(sprintf('%s_%d.mat',dancer{k},j));
21       cS  = size(s.state,2);
22       if( cS > 7 )
23         i          = i + 1;
24         d{i,1}   = s.state; %#ok<*SAGROW>
25         t{i,1}   = s.time;
26         c(i,1)   = k;
27       end
28     end
29   end
30
31   fprintf('%d remaining data sets out of %d total.\n',i,n)
32
33   for k = 1:4
34     j = length(find(c==k));
35     fprintf('%7s data sets %d\n',dancer{k},j)
36   end
37
38   n = i;
39
40   cd ..
41
42   % Limit the range to 6 seconds
43   tRange = 6;
44   for i = 1:n
45     j = find(t{i} - t{i,1} > tRange );
46     if( ~isempty(j) )
47       d{i}(:,j(1)+1:end)= [];
48     end
49   end
```

Four Ryoko sets were lost due to errors in data collection, as shown in the following output:

```
>> DancerNN
 36 remaining data sets out of 40 total.
    Ryoko data sets 6
   Shaye data sets 10
   Emily data sets 10
 Matanya data sets 10
```

We set up and then train the neural network. We use a bidirectional LSTM to classify the sequences. There are ten features: four quaternion measurements, three rate gyro measurements, and three accelerometer measurements. The four quaternion numbers are coupled through the relationship

$$1 = q_1^2 + q_2^2 + q_3^2 + q_4^2 \tag{7.19}$$

However, this should not impact the learning accuracy aside from slowing down the learning.

The bidirectional LSTM is a good choice because we have access to the full sequence. For a classifier using `bilstmLayer`, we must set the `'outputMode'` to `'last'`. This is followed by a fully connected layer, a softmax for producing normalized maximums, and finally the classification layer.

```
50
51   %% Set up the network
52   numFeatures = 10; % 4 quaternion, 3 rate gyros, 3 accelerometers
53   numHiddenUnits = 400;
54   numClasses = 4; % Four dancers
55
56   layers = [ ...
57       sequenceInputLayer(numFeatures)
58       bilstmLayer(numHiddenUnits,'OutputMode','last')
59       fullyConnectedLayer(numClasses)
60       softmaxLayer
61       classificationLayer];
62   disp(layers)
63
64   options = trainingOptions('adam', ...
65       'MaxEpochs',60, ...
66       'GradientThreshold',1, ...
67       'Verbose',0, ...
68       'Plots','training-progress');
69
70   %% Train the network
71   nTrain  = 30;
72   kTrain  = randperm(n,nTrain);
73   xTrain  = d(kTrain);
74   yTrain  = categorical(c(kTrain));
75   net     = trainNetwork(xTrain,yTrain,layers,options);
```

The layers as displayed are shown as follows:

```
5x1 Layer array with layers:

    1   ''   Sequence Input         Sequence input with 10 dimensions
    2   ''   BiLSTM                 BiLSTM with 400 hidden units
    3   ''   Fully Connected        4 fully connected layer
    4   ''   Softmax                softmax
    5   ''   Classification Output  crossentropyex
```

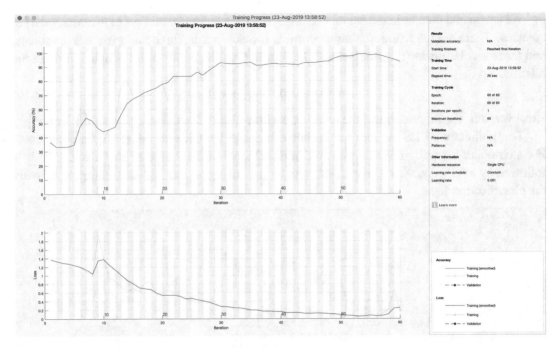

Figure 7.13: *Neural net training.*

The training GUI is shown in Figure 7.13. It converges fairly well.
We test the neural network against the unused data.

```
77  %% Test the network
78  kTest    = setdiff(1:n,kTrain);
79  xTest    = d(kTest);
80  yTest    = categorical(c(kTest));
81  yPred    = classify(net,xTest);
82
83  % Calculate the classification accuracy of the predictions.
84  acc          = sum(yPred == yTest)./numel(yTest);
85  disp('Accuracy')
```

The accuracy achieved is printed as follows:

```
Accuracy
    0.8333
```

This result, > 80%, is pretty good considering the limited amount of data. Interestingly, the deep learning network could distinguish the dancers' pirouettes. The data itself did not show any differences that were easy to identify to the human eye. Calibration could have been done better to make the data more consistent between dancers. It would have been interesting to collect data on multiple days. Other experiments would be to classify pirouettes done in pointe shoes and without. We might also have had the dancers do different types of turns to see if the network could still identify the dancer.

7.10 Data Acquisition GUI

7.10.1 Problem

Build a data acquisition GUI to display the real-time data and output it into training sets.

7.10.2 Solution

Integrate all the preceding recipes into a GUI.

7.10.3 How It Works

We aren't going to use MATLAB's GUIDE or app tools to build our GUI. We will hand-code it, which will give you a better idea of how a GUI works.

We will use nested functions for the GUI. The inner functions have access to all variables in the outer functions. This also makes using callbacks easy as shown in the following code snippet:

```
function DancerGUI( file )
function DrawGUI(h)
  uicontrol( h.fig,'callback',@SetValue);
    function SetValue(hObject, ~, ~ )
    % do something
   end
 end
end
```

A callback is a function called by a `uicontrol` when the user interacts with the control. When you first open the GUI, it will look for the Bluetooth device. This can take a while.

Everything in `DrawGUI` has access to variables in `DancerGUI`. The GUI is shown in Figure 7.14. The 3D orientation display is in the upper-left corner. Real-time plots are on the right. Buttons are on the lower left, and the movie window is on the right.

The upper-left picture shows the dancer's orientation. The plots on the right show angular rates and accelerations from the IMU. From top to bottom of the buttons

1. Turn the 3D on/off. The default model is big, so unless you add your model with fewer vertices, it should be set to off.

2. The text box to its right is the name of the file. The GUI will add a number to the right of the name for each run.

3. Save saves the current data to a file.

4. Calibrate sets the default orientation and sets the gyro rates and accelerations to whatever it is reading when you hit the button. The dancer should be still when you hit calibrate. It will automatically compute the gravitational acceleration and subtract it during the test.

5. Quit closes the GUI.

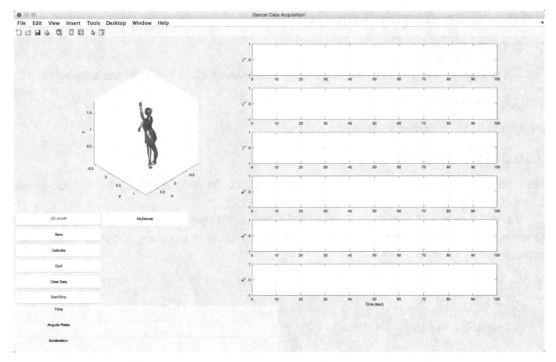

Figure 7.14: *Data acquisition GUI.*

6. Clear data clears out all the internal data storage.

7. Start/Stop starts and stops the GUI.

The remaining three lines display the time, the angular rate vector, and the acceleration vector as numbers. This is the same data that is plotted.

The first part creates the figure and draws the GUI. It initializes all the fields for GUIPlots. It reads in a default picture for the movie window as a placeholder.

DancerGUI.m

```
16  function DancerGUI( file )
17
18  % Demo
19  if( nargin < 1 )
20    DancerGUI('Ballerina.obj');
21    return
22  end
23
24  % Storage of data need by the deep learning system
25  kStore      = 1;
26  accelStore  = zeros(3,1000);
27  gyroStore   = zeros(3,1000);
28  quatStore   = zeros(4,1000);
```

```
29  timeStore    = zeros(1,1000);
30  time         = 0;
31  on3D         = false;
32  quitNow      = false;
33
34  sZ = get(0,'ScreenSize') + [99 99 -200 -200];
35
36  h.fig = figure('name','Dancer Data Acquisition','position',sZ,...
37      'units','pixels','NumberTitle','off','tag','DancerGUI',...
38      'color',[0.9 0.9 0.9]);
39
40  % Plot display
41  gPlot.yLabel = {'\omega_x' '\omega_y' '\omega_z' 'a_x' 'a_y' 'a_z'};
42  gPlot.tLabel = 'Time (sec)';
43  gPlot.tLim   = [0 100];
44  gPlot.pos    = [0.45    0.88    0.46    0.1];
45  gPlot.color  = 'b';
46  gPlot.width  = 1;
47
48  % Calibration
49  q0           = [1;0;0;0];
50  a0           = [0;0;0];
51
52  dIMU.accel = a0;
53  dIMU.quat  = q0;
54
55  % Initialize the GUI
```

The notation

```
'\omega_x'
```

is latex format. This will generate ω_x in the plot labels.

The next part tries to find Bluetooth. It first sees if Bluetooth is available at all. It then enumerates all Bluetooth devices. It looks through the list to find our IMU.

```
57
58  % Get bluetooth information
59  instrreset; % Just in case the IMU wasn't close properly
60  btInfo  = instrhwinfo('Bluetooth');
62
63  if( ~isempty(btInfo.RemoteIDs) )
64      % Display the information about the first device discovered
65      btInfo.RemoteNames(1)
66      btInfo.RemoteIDs(1)
67      for iB = length(btInfo.RemoteIDs)
68          if( strcmp(btInfo.RemoteNames(iB),'LPMSB2-4B31D6') )
69              break;
70          end
71      end
72      b = Bluetooth(btInfo.RemoteIDs{iB}, 1);
```

```
73    fopen(b); % No output allowed for some reason
74    noIMU = false;
75    a     = fread(b,91);
76    dIMU  = DataFromIMU( a );
77  else
78    warndlg('The IMU is not available.', 'Hardware Configuration')
79    noIMU  = true;
```

The following is the run loop. If no IMU is present, it synthesizes data. If the IMU is found, the GUI reads data from the IMU in 91-byte chunks. The `uiwait` is to wait until the user hits the start button. When used for testing, the IMU should be on the dancer. The dancer should remain still when the start button is pushed. It will then calibrate the IMU. Calibration fixes the quaternion reference and removes the gravitational acceleration. You can also hit the calibration button at any time.

```
81
82  % Wait for user input
83  uiwait;
84  % The run loop
85  time = 0;
86  tic
87  while(1)
88    if( noIMU )
89      omegaZ = 2*pi;
90      dT     = toc;
91      time   = time + dT;
92      tic
93      a      = omegaZ*time;
94      q      = [cos(a);0;0;sin(a)];
95      accel  = [0;0;sin(a)];
96      omega  = [0;0;omegaZ];
97    else
98      % Query the bluetooth device
99      a = fread(b,91);
100     pause(0.1); % needed so not to overload the bluetooth device
101
102     dT   = toc;
103     time = time + dT;
104     tic
105
106     % Get a data structure
107     if( length(a) > 1 )
108       dIMU = DataFromIMU( a );
109     end
110     accel  = dIMU.accel - a0;
111     omega  = dIMU.gyro;
112     q      = QuaternionMultiplication(q0,dIMU.quat);
113
114     timeStore(1,kStore)  = time;
115     accelStore(:,kStore) = accel;
116     gyroStore(:,kStore)  = omega;
```

148

```
117    quatStore(:,kStore)   = q;
118    kStore = kStore + 1;
119  end
```

The following closes the GUI. It uses a variable set in one of the callbacks.

```
120
121  if( quitNow )
122    close( h.fig )
```

This code displays the IMU data.

```
123      return
124    else
125      if( on3D )
126        QuaternionVisualization( 'update', q );
127      end
128      set(h.text(1),'string',sprintf('[%5.2f;%5.2f;%5.2f] m/s^2',accel));
129      set(h.text(2),'string',sprintf('[%5.2f;%5.2f;%5.2f] rad/s',omega));
130      set(h.text(3),'string',datestr(now));
131      gPlot = GUIPlots( 'update', [omega;accel], time, gPlot );
132    end
133  end
```

The drawing code uses `uicontrol` to create all the buttons. `GUIPlots` and `Quaterni-onVisualization` are also initialized. The `uicontrol` that require an action have call-backs.

```
135  %% DancerGUI>DrawGUI
136    function DrawGUI
137
138    % Plots
139    gPlot = GUIPlots( 'initialize', [], [], gPlot );
140
141    % Quaternion display
142    subplot('position',[0.05 0.5 0.4 0.4],'DataAspectRatio',[1 1 1],...
143            'PlotBoxAspectRatio',[1 1 1] );
144    QuaternionVisualization( 'initialize', file, h.fig );
145
146    % Buttons
147    f   = {'Acceleration', 'Angular Rates' 'Time'};
148    n   = length(f);
149    p   = get(h.fig,'position');
150    dY  = p(4)/20;
151    yH  = p(4)/21;
152    y   = 0.5;
153    x   = 0.15;
154    wX  = p(3)/6;
155
156    % Create pushbuttons and defaults
157    for k = 1:n
```

```
158    h.pushbutton(k) = uicontrol( h.fig,'style','text','string',f{k},...
159                        'position',[x y wX yH]);
160    h.text(k) = uicontrol( h.fig,'style','text','string','',...
161                        'position',[x+wX y 2*wX yH]);
162    y = y + dY;
163  end
164
165  h.onButton = uicontrol( h.fig,'style','togglebutton',...
166    'string','Start/Stop','position',[x y wX yH],'ForegroundColor','red
        ',...
167    'callback',@StartStop);
168  y = y + dY;
169  h.clrButton = uicontrol( h.fig,'style','pushbutton',...
170    'string','Clear Data','position',[x y wX yH],'callback',@Clear);
171  y = y + dY;
172  h.quitButton = uicontrol( h.fig,'style','pushbutton',...
173    'string','Quit','position',[x y wX yH],'callback',@Quit);
174  y = y + dY;
175  h.calibrateButton = uicontrol( h.fig,'style','pushbutton',...
176    'string','Calibrate','position',[x y wX yH],'callback',@Calibrate);
177  y = y + dY;
178  h.saveButton = uicontrol( h.fig,'style','pushbutton','string','Save'
        ,...
179    'position',[x y wX yH],'callback',@SaveFile);   y = y + dY;
180  h.on3D = uicontrol( h.fig,'style','togglebutton','string','3D on/off'
        ,...
181    'position',[x y wX yH],'ForegroundColor','red','callback',@On3D);
182
183  h.matFile = uicontrol( h.fig,'style','edit', 'string','MyDancer',...
184    'position',[x+wX y wX yH]);
```

`uicontrol` takes parameter pairs, except for the first argument that can be a figure handle. There are a lot of parameter pairs. The easiest way to explore them is to type

```
h = uicontrol;
get(h)
```

All types of `uicontrol` that handle user interaction have "callbacks" that are functions that do something when the button is pushed or a menu item is selected. We have five `uicontrol` with callbacks. The start/stop button callback uses `uiwait` and `uiresume` to start and stop data collection.

```
191    % Start/Stop button callback
192    function StartStop(hObject, ~, ~ )
193      if( hObject.Value )
194        uiresume;
195      else
196        SaveFile;
197        uiwait
198      end
199    end
```

Figure 7.15: *Modal dialog.*

The Quit button uses `questdlg` to ask if you want to save the data that has been stored in the GUI. This produces the modal dialog shown in Figure 7.15.

```matlab
201  % Quit button callback
202    function Quit(~, ~, ~ )
203      button = questdlg('Save Data?','Exit Dialog','Yes','No','No');
204      switch button
205        case 'Yes'
206           % Save data
207        case 'No'
208      end
209      quitNow = true;
210      uiresume
211    end
```

The `Clear` callback function clears the data storage arrays. It resets the quaternion to a unit quaternion.

```matlab
213  % Clear button callback
214    function Clear(~, ~, ~ )
215      kStore      = 1;
216      accelStore  = zeros(3,1000);
217      gyroStore   = zeros(3,1000);
218      quatStore   = zeros(4,1000);
219      timeStore   = zeros(1,1000);
220      time        = 0;
221    end
```

The `calibrate` callback runs the calibration procedure.

```matlab
223  % Calibrate button callback
224    function Calibrate(~, ~, ~ )
225      a         = fread(b,91);
226      dIMU      = DataFromIMU( a );
227      a0        = dIMU.accel;
228      q0        = dIMU.quat;
229      QuaternionVisualization( 'update', q0 )
230    end
```

The last, `SaveFile`, saves the recorded data into a mat-file for use by the deep learning algorithm.

```
232   % Save button call back
233   function SaveFile(~,~,~)
234     cd TestData
235     fileName = get(h.matFile,'string');
236     s = dir;
237     n = length(s);
238     fNames = cell(1,n-2);
239     for kF = 3:n
240       fNames{kF-2} = s(kF).name(1:end-4);
241     end
242     j = contains(fNames,fileName);
243     m = 0;
244     if( ~isempty(j) )
245       for kF = 1:length(j)
246         if( j(kF))
247           f = fNames{kF};
248           i = strfind(f,'_');
249           m = str2double(f(i+1:end));
250         end
251       end
252     end

254     timeStore(kStore:end) = [];
255     gyroStore(:,kStore:end) = [];
256     quatStore(:,kStore:end) = [];
257     accelStore(:,kStore:end) = [];

259     state = [gyroStore;accelStore;quatStore];
260     time  = timeStore;

262     save(sprintf('%s_%d.mat',fileName,m+1), 'state','time');

264     kStore      = 1;
265     accelStore  = zeros(3,1000);
266     gyroStore   = zeros(3,1000);
267     quatStore   = zeros(4,1000);
268     timeStore   = zeros(1,1000);
269     time        = 0;

271     cd ..

273     gPlot = GUIPlots( 'initialize', [], [], gPlot );

275   end
```

We make it easier for the user to save files by reading the directory and adding a number to the end of the dancer filename that is one greater than the last filename number.

7.11   Hardware Sources

Table 7.3 gives the hardware used in this chapter along with the prices (in US dollars) at the time of publication.

Table 7.3: Hardware.

Component	Supplier	Part Number	Price
IMU	LP-Research Inc.	LPMS-B2: 9-Axis Inertial Measurement Unit	$299.00
IMU Holder	LP-Research Inc.	LPMS-B2: Holder	$30.00
Belt	Amazon	Men's Elastic Stretch Belt Invisible Casual Trousers Webbing Belt Plastic Buckle Black Fits 24" to 42"	$10.99

CHAPTER 8

■ ■ ■

Completing Sentences

8.1   Introduction

8.1.1   Sentence Completion

Completing sentences is a useful feature for text entry systems. Given a set of possible sentences, we want the system to predict a missing part of the sentence. We will use the Research Sentence Completion Challenge [46]. It is a database of 1040 sentences each of which has four imposter sentences and one correct sentence. Each imposter sentence differs from the correct sentence by one word in a fixed position. The deep learning system should identify the correct word in the sentence. Imposter words have similar occurrence statistics. The sentences were selected from Sherlock Holmes novels. The imposter words were generated using a language model trained using over 500 nineteenth-century novels. Thirty alternative words for the correct word were produced. Human judges picked the four best imposter words from the 30 alternatives. The database can be downloaded from Google Drive [27].

The first question in the database and the five answers, including the four imposters, are given as follows:

```
I have it from the same source that you are both an orphan and a bachelor
   and are _____ alone in London.
 a) crying  b)  instantaneously c) residing d) matched e) walking
```

(b) and (d) don't fit grammatically. (a) and (e) are incompatible with the beginning in which the speaker is recounting general information about the subject's state. (c) makes the most sense. If after "are" we had "often seen," then (a) and (e) would be possibilities, and (c) would no longer make sense. You would need additional information to determine if (a) or (e) were correct.

The first few recipes in this chapter are dedicated to preparing the data from our online source. The final recipe, 8.6, creates and trains a deep learning net to complete sentences.

© Michael Paluszek, Stephanie Thomas, Eric Ham 2022
M. Paluszek et al., *Practical MATLAB Deep Learning*,
https://doi.org/10.1007/978-1-4842-7912-0_8

8.1.2 Grammar

Grammar is important in interpreting sentences. The structure of a language, that is, its grammar, is very important. Since not all of our readers speak English as their primary language, we'll give some examples in other languages.

In Russian, the word order is not fixed. You can always figure out the words and whether they are adjectives, verbs, nouns, and so forth from the declension and conjugation, but the word order is important as it determines the emphasis. For example, to say, "I am an engineer" in Russian:

$$Я \ инженер$$

We could reverse the order

$$инженер \ Я$$

which would mean the emphasis is on "engineer" not "I." While it is easy to know that the sentence is stating that "I am an engineer," we don't necessarily know how the speaker feels about it. This may not be important in rote translation but certainly makes a difference in literature.

Japanese is known as a subject-object-verb language. In Japanese, the verb is at the end of the sentence. Japanese also makes use of particles to denote word function such as subject or object. For the sentence completion problem, the particle would denote the function of the word. The rest of the sentence would determine what the word may mean. Here are some particles:

は"wa/ha" indicates the topic, which could be the object or subject.
を"wo/o" indicates the object.
が "ga" indicates the subject.

For example, in Japanese

$$私はエンジニアです$$

or "watashi wa enjinia desu"
means "I am an engineer." はis the topic marker pointing to "I." ですis the verb. We'd need other sentences to predict the 私, "I," or "engineer."
Japanese also has the feature where everything, except the verb, can be omitted.

$$いただきます$$

or "I ta da ki ma su." This means "I will eat" whatever given. You need to know the context or have other sentences to understand what is meant by the sentence.

In addition, in Japanese, many different Kanji, or symbols, can mean approximately the same thing, but the emphasis will be different. Other Kanji have different meanings depending on the context. Japanese also does not have any spaces between words. You just have to know when a kana character, like は, is part of the preceding Kanji. For example, verb conjugation uses Hiragana characters to indicate past, present, etc. For example:

$$食べる$$

is "to eat." There is a Kanji root and then two Hiragana to form the entire verb. By itself, it is a legitimate sentence, and you need the context to determine who is eating. The negative verb, "not eat," is

$$食べない$$

with the Hiragana

$$ない$$

replacing

$$る$$

to form the negative.

Every language needs to be approached a little differently when you are trying to do natural language processing.

8.1.3 Sentence Completion by Pattern Recognition

Our approach is sentence completion by pattern recognition. Given a database of your sentences, the pattern recognition algorithm should be able to recognize the patterns you use and find errors. Also, in most languages, dialogs between people use far fewer words and simpler structures than the written languages. You will notice this if you watch a movie in a foreign language for which you have a passable knowledge. You can recognize a lot more than you would expect. Russian is an extreme in this regard; it is very hard to build vocabulary from reading because the language is so complex. Many Russian teachers teach the root system so that you can guess word meaning without constantly referring to a dictionary. Using word roots and sentence structure to guess words is a form of sentence completion. We'll leave that to our Russian readers.

8.1.4 Sentence Generation

As an aside, sentence completion leads to generative deep learning [13]. In generative deep learning, the neural network learns patterns and then can create new material. For example, a deep learning network might learn how a newspaper article is written and be able to generate new articles given basic facts the article is supposed to present. This is not a whole lot different than when writers are paid to write new books in a series such as *Tom Swift* or *Nancy Drew*. Presumably, the writer adds his or her personality to the story, but perhaps a reader, who just wants a page-turner, wouldn't really care.

8.2   Generating a Database

8.2.1 Problem

We want to create a set of sentences accessible from MATLAB.

8.2.2 Solution

Read in the sentences from the database. Write a function, `ReadDatabase.m`, to read in tab-separated text.

8.2.3 How It Works

The database that we downloaded from Google Drive was an Excel `csv` file. We need to first open the file and save it as tab-delimited text. Once this is done, you are ready to read it into MATLAB. We do this for both `test_answer.csv` and `testing_data.csv`. We manually removed the first column in `test_answer.csv` in Excel because it was not needed. Only the `txt` files that we generated are needed in this book.

If you have the Statistics and Machine Learning Toolbox, you could use `tdfread - s = tdfread(file,delimiter)`. We'll write the equivalent. There are three outputs shown in the header. They are the sentences, the range of characters where the word needed for completion fits, the five possible words, and the answer.

We open the file using `f = fopen('testing_data.txt','r');`. This tells it that the file is a text file. We search for tabs and add the end of the line so that we can find the last word. The second read reads in the test answers and converts them from a character to a number. We removed all extraneous quotes from the text file with a text editor.

ReadDatabase.m

```
18  function [s,u,v,a] = ReadDatabase
19
20  f = fopen('testing_data.txt','r');
21
22  % We know the size of the file simplying the code.
23  u = zeros(1040,2);
24  a = zeros(1040,1);
25  s = strings(1040,1);
26  v = strings(1040,5);
27  t = sprintf('\t');
28  k = 1;
29
30  % Read in the sentences and words
31  while(~feof(f))
32      q       = fgetl(f); % This is one line of text
33      j       = [strfind(q,t) length(q)+1]; % This finds tabs that delimit
                words
```

158

```
34    s(k)    = convertCharsToStrings(q(j(1)+1:j(2)-1)); % Convert to strings
35    for i = 1:5
36      v(k,i) = convertCharsToStrings(q(j(i+1)+1:j(i+2)-1)); % Make
            strings
37    end
38    ul      = strfind(s(k),'_'); % Find the space where the answers go
39    u(k,:)  = [ul(1) ul(end)]; % Get the range of characters for the
          answer
40    k       = k + 1;
41  end
42
43  fclose(f);
44
45  % Read in the test answers
46  f = fopen('test_answer.txt','r');
47
48  k = 1;
49  while(~feof(f))
50    q             = fgetl(f);
51    a(k,1)        = double(q)-96;
52    k             = k + 1;
53  end
54
55  fclose(f);
```

If we run the function, we get the following outputs:

```
>> [s,u,v,a] = ReadDatabase;
>> s(1)
ans =
    "I have it from the same source that you are both an orphan and a
        bachelor and are _____ alone in London."
>> v(1,:)
ans =
  1x5 string array
    "crying"    "instantaneously"    "residing"    "matched"    "walking"
>> a(1)
ans =
    3
```

All outputs (except for the answer number) are strings. convertCharsToStrings does the conversion. Now that we have all of the data in MATLAB, we are ready to train the deep learning system to determine the best word for each sentence. As an intermediate step, we will convert the words to numbers.

8.3 Creating a Numeric Dictionary

8.3.1 Problem

We want to create a numeric dictionary to speed neural net training. This eliminates the need for string matching during the training process. Expressing a sentence as a numeric sequence as opposed to a sequence of character arrays (words) essentially gives us a more efficient way to represent the sentence. This will become useful later when we perform machine learning over a database of sentences to learn valid and invalid sequences.

8.3.2 Solution

Write a MATLAB function to search through text and find unique words.

8.3.3 How It Works

The function removes punctuation using `erase` in the following lines of code.

DistinctWords.m

```matlab
16  function [d,n] = DistinctWords( w )
17
18  % Demo
19  if( nargin < 1 )
20    Demo;
21    return
22  end
23
24  % Remove punctuation
25  w = erase(w,';');
26  w = erase(w,',');
27  w = erase(w,'.');
```

It then uses `split` to break up the string and finds unique strings using `unique`.

```matlab
29  % Find unique words
30  s = split(w)';
31  d = unique(s);
32
33  if( nargout > 1 )
34    % find the numbers for the words
35    n = zeros(1,length(s));
36    for k = 1:length(s)
37      for j = 1:length(d)
38        if( d(j) == s(k) )
39          n(k) = j;
40          break;
41        end
42      end
43    end
44  end
```

This is the built-in demo. It finds 38 unique words.

```
>> DistinctWords
w =
    "No one knew it then, but she was being held under a type of house
        arrest while the tax authorities scoured
    the records of her long and lucrative career as an actress, a
        luminary of the red carpet, a face of luxury
    brands and a successful businesswoman."
d =
    1x38 string array
    Columns 1 through 12
      "No"     "one"     "knew"     "it"     "then"     "but"     "she"     "was"
            "being"     "held"     "under"     "type"
    Columns 13 through 22
      "house"     "arrest"     "while"     "tax"     "authorities"     "scoured"
            "records"     "her"     "long"     "lucrative"
    Columns 23 through 33
      "career"     "as"     "an"     "actress"     "luminary"     "the"     "red"
            "carpet"     "face"     "of"     "luxury"
    Columns 34 through 38
      "brands"     "and"     "a"     "successful"     "businesswoman"
n =
    Columns 1 through 20
      1     2     3     4     5     6     7     8     9    10    11    36
           12    32    13    14    15    28    16    17
    Columns 21 through 40
     18    28    19    32    20    21    35    22    23    24    25    26
           36    27    32    28    29    30    36    31
    Columns 41 through 47
     32    33    34    35    36    37    38
```

d is a string array and maps onto array n.

8.4   Mapping Sentences to Numbers

8.4.1 Problem

We want to map words in sentences to unique numbers.

8.4.2 Solution

Write a MATLAB function to search through text and assign a unique number to each word. This approach will have problems with homonyms.

8.4.3 How It Works

The function splits the string and searches using d. The last line removes any words (in this case, only punctuation) that are not in the dictionary.

MapToNumbers.m

```matlab
18  function n = MapToNumbers( w, d )
19
20  % Demo
21  if( nargin < 1 )
22    Demo;
23    return
24  end
25
26  w = erase(w,';');
27  w = erase(w,',');
28  w = erase(w,'.');
29  s = split(w)';  % string array
30
31  n = zeros(1,length(s));
32  for k = 1:length(s)
33    ids = find(strcmp(s(k),d));
34    if ~isempty(ids)
35      n(k) = ids;
36    end
37
38  end
39
40  n(n==0) = [];
```

This is the built-in demo.

```
>> MapToNumbers
w =
    "No one knew it then, but she was being held under a type of house
     arrest while the tax authorities scoured the records of her long
     and lucrative career as an actress, a luminary of the red carpet,
     a face of luxury brands and a successful businesswoman."
n =
  Columns 1 through 19
     1     2     3     4     0     6     7     8     9    10    11    36
          12    32    13    14    15    28    16
  Columns 20 through 38
    17    18    28    19    32    20    21    35    22    23    24    25
           0    36    27    32    28    29     0
  Columns 39 through 46
    36    31    32    33    34    35    36    37
```

8.5   Converting the Sentences

8.5.1 Problem

We want to convert the sentences to numeric sequences.

8.5.2 Solution

Write a MATLAB script to take each sentence in our database, add the words, and create a sequence. Each sentence is classified as correct or incorrect.

8.5.3 How It Works

The script reads in the database. It creates a numeric dictionary for all of the sentences and then converts them to numbers. The numeric data is then saved to a mat-file for easy access later. This first part of the script creates 5200 sentences. Each sentence is classified as correct or incorrect. Note how we initialize a string array.

PrepareSequences.m

```
1   %% Script to create all the sentences and labels
5
6   [s,u,v,a] = ReadDatabase;
7
8   %    s {1040,1} Sentences
9   %    u (1040,2) Range of underlines in sentence
10  %    v {1040,5} Possible words including 4 imposters
11  %    a (1040,1) Which of the five words is correct
13
14  % Whatever you want in the training, ex. 100, length(s) for all.
15  nSentences = length(s);
16
17  i = 1;
18  c = zeros(size(v,2)*nSentences,1);
19  z = strings(size(v,2)*nSentences,1);
20  % Extract the parts of the sentences before and after the underlined
        part
21  % Then create sentences with all possible words
22  for k = 1:nSentences
23      q1    = extractBefore(s(k),u(k,1));
24      q2    = extractAfter(s(k),u(k,2));
25      for j = 1:size(v,2)
26          z(i)  = q1 + v(k,j) + q2;
27          if( j == a(k,1) )
28              c(i) = 1;
29          else
30              c(i) = 0;
31          end
32          i = i + 1;
33      end
34  end
```

The next section concatenates all of the sentences into a gigantic string and creates a dictionary.

```
37  r = z(1);
38  for k = 2:length(z)
39    r = r + " " + z(k); % append all the sentences to one string
40  end
```

The final part creates the numeric sentences and saves them. This part uses `MapTo Numbers` from the previous recipe. The loop that prints the lines shows a handy way of printing an array using `fprintf`.

```
42  d = DistinctWords( r ); % find the distinct words
43
44  nZ = cell(length(z),1);
45  for k = 1:length(z)
46    nZ{k} = MapToNumbers( z(k), d );
47  end
```

As expected, only one word is different in each set of five sentences.

```
>> PrepareSequences
Category: 0    1 428 538 541 553 103    6 535 149  10    7 170    8 546 544    9 546  10    2  12 404
Category: 0    1 428 538 541 553 103    6 535 149  10    7 170    8 546 544    9 546  10    3  12 404
Category: 1    1 428 538 541 553 103    6 535 149  10    7 170    8 546 544    9 546  10    4  12 404
Category: 0    1 428 538 541 553 103    6 535 149  10    7 170    8 546 544    9 546  10    5  12 404
Category: 0    1 428 538 541 553 103    6 535 149  10    7 170    8 546 544    9 546  10   11  12 404

Category: 1 323 481 378  19 465 544  18 546  19 465 544  20 549  21  22  14 404  24  25 546
Category: 0 323 481 378  19 465 544  18 546  19 465 544  20 549  21  22  15 404  24  25 546
Category: 0 323 481 378  19 465 544  18 546  19 465 544  20 549  21  22  16 404  24  25 546
Category: 0 323 481 378  19 465 544  18 546  19 465 544  20 549  21  22  17 404  24  25 546
Category: 0 323 481 378  19 465 544  18 546  19 465 544  20 549  21  22  23 404  24  25 546
```

8.6 Training and Testing

8.6.1 Problem

We want to build a deep learning system to complete sentences. The idea is that the full database of correct and incorrect sentences provides enough information for the neural net to determine which word is the correct word for the sentence.

8.6.2 Solution

Write a MATLAB script to implement an LSTM to classify the sentences as correct or incorrect. The LSTM will be trained with complete sentences. No information about the words, such as whether a word is noun, verb, or adjective, nor of any grammatical structure will be used.

8.6.3 How It Works

We will produce the simplest possible design. It will read in sentences, classified as correct or incorrect, and attempt to determine if new sentences are correct or incorrect just from the learned patterns. This is a very simple and crude approach. We aren't taking advantage of our knowledge of grammar, word types (verb, noun, etc.), or context to help with the predictions. Language modeling is a huge field, and we are not using any results from that body of work. Of course, applications of all of the rules of grammar don't necessarily ensure success; otherwise, there would be more 800s on the SAT verbal tests. We'll show two different sets of layers. The first will produce a good fit, and the second will overfit.

We use the same code as the previous recipe to make sure the sequences are valid. We use a `clear all` so that the sentences are always the same.

SentenceCompletionNNFitted.m

```
7   clear all
8
9   %% Load the data
10  s = load('Sentences');
11  n = length(s.c);          % number of sentences
12
13  % Make sure the sequences are valid. One in every 5 is complete.
14  for k = 1:10
15    fprintf('Category: %d',s.c(k));
16    fprintf('%5d',s.nZ{k})
17    fprintf('\n')
18    if( mod(k,5) == 0 )
19      fprintf('\n')
20    end
```

The layers were designed to get a well-fitted training. As the training plot shows, the validation error is the same as the training error. Because we have access to full sequences at prediction time, we use a bidirectional LSTM layer in the network. A bidirectional LSTM layer learns from the full sequence at each step. We use three `BiLSTM` layers with fully connected layers and dropouts in between.

SentenceCompletionNNFitted.m

```
22
23  %% Set up the network
24  numFeatures = 1;
25  numClasses = 2;
26
27  %
28  layers = [ ...
29      sequenceInputLayer(numFeatures)
30      bilstmLayer(80,'OutputMode','sequence')
31      fullyConnectedLayer(numClasses)
32      dropoutLayer(0.4)
```

```
33      bilstmLayer(60,'OutputMode','sequence')
34      fullyConnectedLayer(numClasses)
35      dropoutLayer(0.2) % 0.2 was pretty good
36      bilstmLayer(20,'OutputMode','last')
37      fullyConnectedLayer(numClasses)
38      softmaxLayer
39      classificationLayer];
40  disp(layers)
41
42  %% Train the network
43  kTrain      = randperm(n,0.85*n);
44  xTrain      = s.nZ(kTrain);           % sentence indices, in order
45  yTrain      = categorical(s.c(kTrain)); % complete or not?
46
47  % Test this network
48  kTest       = setdiff(1:n,kTrain);
49  xTest       = s.nZ(kTest);
50  yTest       = categorical(s.c(kTest));
51
52  options = trainingOptions('adam', ...
53      'MaxEpochs',60, ...
54      'GradientThreshold',1, ...
55      'ValidationData',{xTest,yTest}, ...
56      'ValidationFrequency',10, ...
57      'Verbose',0, ...
58      'Plots','training-progress');
59
60  disp(options)
61
62  net         = trainNetwork(xTrain,yTrain,layers,options);
63  yPred       = classify(net,xTest);
64
65  % Calculate the classification accuracy of the predictions.
66  acc         = sum(yPred == yTest)./numel(yTest);
67  disp('All')
68  disp(acc);
69
70  j           = find(yTest == '1');
```

The second test tests only correct sentences. It never identifies any sentences as correct. The output of this section is

```
>> SentenceCompletionNNFitted
Category: 0   17   480   551   429   985   845   924   984  1123    91   165    79   713    81
          47   116    81    91   286    76   524    23
Category: 0   17   480   551   429   985   845   924   984  1123    91   165    79   713    81
          47   116    81    91   539    76   524    23
Category: 1   17   480   551   429   985   845   924   984  1123    91   165    79   713    81
          47   116    81    91   817    76   524    23
Category: 0   17   480   551   429   985   845   924   984  1123    91   165    79   713    81
          47   116    81    91   621    76   524    23
Category: 0   17   480   551   429   985   845   924   984  1123    91   165    79   713    81
          47   116    81    91  1066    76   524    23
```

```
Category: 1   19 1070  436  740   99  47  898  81  740   99  47  133 1103  410
        95  291  524  371  679   81  264
Category: 0   19 1070  436  740   99  47  898  81  740   99  47  133 1103  410
        95  776  524  371  679   81  264
Category: 0   19 1070  436  740   99  47  898  81  740   99  47  133 1103  410
        95  525  524  371  679   81  264
Category: 0   19 1070  436  740   99  47  898  81  740   99  47  133 1103  410
        95  644  524  371  679   81  264
Category: 0   19 1070  436  740   99  47  898  81  740   99  47  133 1103  410
        95  236  524  371  679   81  264


11x1 Layer array with layers:

    1    ''   Sequence Input          Sequence input with 1 dimensions
    2    ''   BiLSTM                  BiLSTM with 80 hidden units
    3    ''   Fully Connected         2 fully connected layer
    4    ''   Dropout                 40% dropout
    5    ''   BiLSTM                  BiLSTM with 60 hidden units
    6    ''   Fully Connected         2 fully connected layer
    7    ''   Dropout                 20% dropout
    8    ''   BiLSTM                  BiLSTM with 20 hidden units
    9    ''   Fully Connected         2 fully connected layer
   10    ''   Softmax                 softmax
   11    ''   Classification Output   crossentropyex
TrainingOptionsADAM with properties:

          GradientDecayFactor: 0.9000
   SquaredGradientDecayFactor: 0.9990
                      Epsilon: 1.0000e-08
              InitialLearnRate: 1.0000e-03
     LearnRateScheduleSettings: [1x1 struct]
              L2Regularization: 1.0000e-04
        GradientThresholdMethod: 'l2norm'
             GradientThreshold: 1
                     MaxEpochs: 60
                  MiniBatchSize: 128
                       Verbose: 0
              VerboseFrequency: 50
                ValidationData: {{75x1 cell}   [75x1 categorical]}
           ValidationFrequency: 10
            ValidationPatience: Inf
                       Shuffle: 'once'
                CheckpointPath: ''
          ExecutionEnvironment: 'auto'
                    WorkerLoad: []
                     OutputFcn: []
                         Plots: 'training-progress'
                SequenceLength: 'longest'
          SequencePaddingValue: 0
          DispatchInBackground: 0
All
    0.7867
Correct
    0
```

The first layer says the input is a one-dimensional sequence. The second is the bidirectional LSTM. The next layer is a fully connected layer of neurons. This is repeated with a dropout

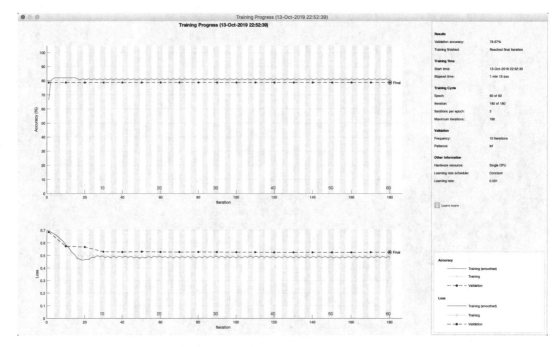

Figure 8.1: *Training progress.*

layer in between. Figure 8.1 shows the training progress. This is followed by a softmax layer and then by the classification layer. The standard softmax is

$$\sigma_k = \frac{e^{z_i}}{\sum_{j=1}^{K} e^{z_j}} \tag{8.1}$$

which is essentially a normalized output.

The testing code is

```
53        'Verbose',0, ...
54        'Plots','training-progress');
55
56  disp(options)
57
58  net          = trainNetwork(xTrain,yTrain,layers,options);
59  yPred        = classify(net,xTest);
60
61  % Calculate the classification accuracy of the predictions.
62  acc          = sum(yPred == yTest)./numel(yTest);
```

The results are not as good as just guessing all the sentences are wrong. When we test the neural net, it never thinks any sentences are correct. This really doesn't solve our problem. Given that only 20% sentences are correct, the neural net scores 80% by saying they are all incorrect.

Our second neural net has a different structure. We have two BiLSTM layers without the fully connected layer in between. Both biLSTM have the same number of hidden units. This was found after trying many different combinations.

The training code follows. We convert the classes, 0 and 1, to a categorical variable.

SentenceCompletionNN.m

```matlab
21  %% Set up the network
22  numFeatures = 1;
23  numHiddenUnits = 200;
24
25  numClasses = 2;
26
27  % Good results with validation frequency of 10 and 200 hidden units
28  layers = [ ...
29      sequenceInputLayer(numFeatures)
30      bilstmLayer(numHiddenUnits,'OutputMode','sequence')
31      dropoutLayer(0.2)
32      bilstmLayer(numHiddenUnits,'OutputMode','last')
33      fullyConnectedLayer(numClasses)
34      softmaxLayer
35      classificationLayer];
36  disp(layers)
37
38  %% Train the network
39  kTrain      = randperm(n,0.85*n);
40  xTrain      = s.nZ(kTrain);             % sentence indices, in order
41  yTrain      = categorical(s.c(kTrain)); % complete or not?
42
43  % Test this network - results show overfitting
44  kTest       = setdiff(1:n,kTrain);
45  xTest       = s.nZ(kTest);
46  yTest       = categorical(s.c(kTest));
47
48  options = trainingOptions('adam', ...
49      'MaxEpochs',240, ...
50      'GradientThreshold',1, ...
51      'ValidationData',{xTest,yTest}, ...
52      'ValidationFrequency',10, ...
53      'Verbose',0, ...
54      'Plots','training-progress');
55
56  disp(options)
57
58  net         = trainNetwork(xTrain,yTrain,layers,options);
59  yPred       = classify(net,xTest);
60
61  % Calculate the classification accuracy of the predictions.
62  acc         = sum(yPred == yTest)./numel(yTest);
63  disp('All')
64  disp(acc);
```

The output is

```
>> SentenceCompletionNN
Category:  0   17  480  551  429  985  845  924  984 1123   91  165   79  713   81
          47  116   81   91  286   76  524   23
Category:  0   17  480  551  429  985  845  924  984 1123   91  165   79  713   81
          47  116   81   91  539   76  524   23
Category:  1   17  480  551  429  985  845  924  984 1123   91  165   79  713   81
          47  116   81   91  817   76  524   23
Category:  0   17  480  551  429  985  845  924  984 1123   91  165   79  713   81
          47  116   81   91  621   76  524   23
Category:  0   17  480  551  429  985  845  924  984 1123   91  165   79  713   81
          47  116   81   91 1066   76  524   23

Category:  1   19 1070  436  740   99   47  898   81  740   99   47  133 1103  410
          95  291  524  371  679   81  264
Category:  0   19 1070  436  740   99   47  898   81  740   99   47  133 1103  410
          95  776  524  371  679   81  264
Category:  0   19 1070  436  740   99   47  898   81  740   99   47  133 1103  410
          95  525  524  371  679   81  264
Category:  0   19 1070  436  740   99   47  898   81  740   99   47  133 1103  410
          95  644  524  371  679   81  264
Category:  0   19 1070  436  740   99   47  898   81  740   99   47  133 1103  410
          95  236  524  371  679   81  264


7x1 Layer array with layers:

    1    ''    Sequence Input          Sequence input with 1 dimensions
    2    ''    BiLSTM                  BiLSTM with 200 hidden units
    3    ''    Dropout                 20% dropout
    4    ''    BiLSTM                  BiLSTM with 200 hidden units
    5    ''    Fully Connected         2 fully connected layer
    6    ''    Softmax                 softmax
    7    ''    Classification Output   crossentropyex
TrainingOptionsADAM with properties:

              GradientDecayFactor: 0.9000
       SquaredGradientDecayFactor: 0.9990
                          Epsilon: 1.0000e-08
                 InitialLearnRate: 1.0000e-03
         LearnRateScheduleSettings: [1x1 struct]
                 L2Regularization: 1.0000e-04
           GradientThresholdMethod: 'l2norm'
                 GradientThreshold: 1
                        MaxEpochs: 240
                    MiniBatchSize: 128
                          Verbose: 0
                 VerboseFrequency: 50
                   ValidationData: {{75x1 cell}  [75x1 categorical]}
              ValidationFrequency: 10
               ValidationPatience: Inf
                          Shuffle: 'once'
                   CheckpointPath: ''
             ExecutionEnvironment: 'auto'
                       WorkerLoad: []
                        OutputFcn: []
                            Plots: 'training-progress'
                   SequenceLength: 'longest'
             SequencePaddingValue: 0
             DispatchInBackground: 0
```

```
All
    0.7333
Correct
    0.1250
```

The testing code is shown as follows:

```
66  j          = find(yTest == '1');
67  yPredC     = classify(net,xTest(j));
68  accC       = sum(yPredC == yTest(j))./numel(yTest(j));
69  disp('Correct')
70  disp(accC);
73
74  %% Copyright
75  %    Copyright (c) 2019 Princeton Satellite Systems, Inc.
```

Figure 8.2 shows training. As the training progresses, the validation loss continues. Because it is higher than the training loss, we see that we are overfitting, that is, we have too many neurons. Our training accuracy is over 73%. The training loss continues to improve, and the validation accuracy gets worse. The accuracy is worse than the case with the good fitting. However, this network will classify sentences as correct, instead of just declaring them all wrong and getting the default 80% accuracy. In that way, it is an improvement though you still wouldn't want to use it as an SAT aid.

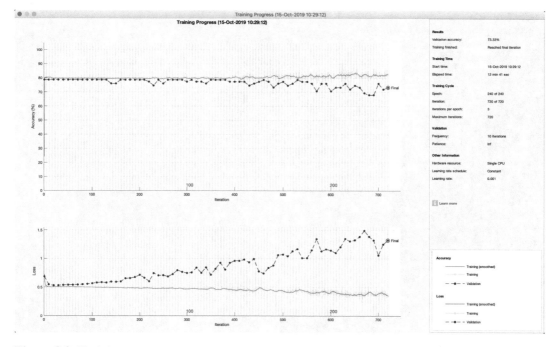

Figure 8.2: *Training progress.*

This network works better than the previous one, despite overfitting. It thinks some sentences are correct and gets that right 12.5% of the time. This may indicate that we do not have enough data for the neural network to form a grammar. The poor performance also helps to show that NLP is a tough problem with many facets of research. You would expect to do better if you tried to take advantage of the grammatical structure of sentences, classifying words into different types, etc.

We've shown that we can get a neural network to be somewhat successful in sentence completion. The problem the training faces is that it can be 80% successful by saying there are no correct fits. This simple approach might work better if we had a larger database of sentences. We didn't use all of the sentences that are available with the book so you can try expanding the set. With enough examples, the network might begin to learn the grammar. It would also be interesting to try this with different languages.

CHAPTER 9

■ ■ ■

Terrain-Based Navigation

9.1 Introduction

Before the widespread availability of GPS, Loran, and other electronic navigation aids, pilots used visual cues from terrain to navigate. Now everyone uses GPS. We want to return to the good old days of terrain-based navigation. We will design a system that will be able to match terrain with a database. It will then use that information to determine where it is flying.

9.2 Modeling Our Aircraft

9.2.1 Problem

We want a three-dimensional aircraft model that can change direction.

9.2.2 Solution

Write the equations of motion for three-dimensional flight.

9.2.3 How It Works

The motion of a point mass through three-dimensional space has three degrees of freedom. Our aircraft model is therefore given three degrees of spatial freedom. The velocity vector is expressed as a wind-relative magnitude (V) with directional components for heading (ψ) and flight path angle (γ). The position is a direct integral of the velocity and is expressed in $y =$ North, $x =$ East, $h =$ Vertical coordinates. In addition, the engine thrust is modeled as a first-order system where the time constant can be changed to approximate the engine response times of different aircraft.

Figure 9.1 shows a diagram of the velocity vector in the North-East-Up coordinate system. The time derivatives are taken in this frame. This is not a purely inertial coordinate system, because it is rotating with the Earth. However, the rate of rotation of the Earth is sufficiently small compared to the aircraft turning rates so that it can be safely neglected.

© Michael Paluszek, Stephanie Thomas, Eric Ham 2022
M. Paluszek et al., *Practical MATLAB Deep Learning*,
https://doi.org/10.1007/978-1-4842-7912-0_9

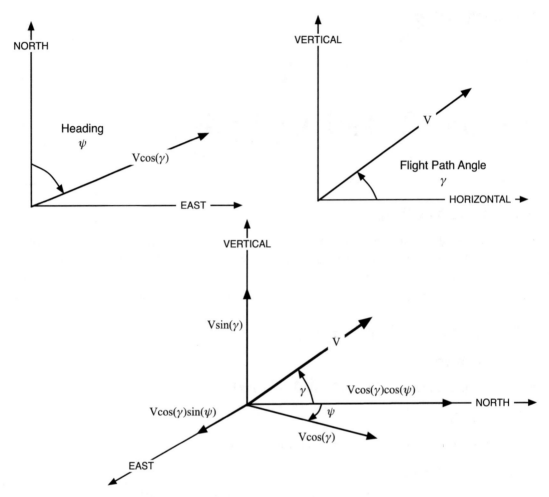

Figure 9.1: *Velocity in North-East-Up coordinates.*

The point-mass aircraft equations of motion are

$$\dot{v} = (T\cos\alpha - D - mg\sin\gamma)/m - f_v \tag{9.1}$$

$$\dot{\gamma} = \frac{1}{mv}\left((L + T\sin\alpha)\cos\phi - mg\cos\gamma + f_\gamma\right) \tag{9.2}$$

$$\dot{\psi} = \frac{1}{mv\cos\gamma}\left((L + T\sin\alpha)\sin\phi - f_\psi\right) \tag{9.3}$$

$$\dot{x}_e = v\cos\gamma\sin\psi + W_x \tag{9.4}$$

$$\dot{y}_n = v\cos\gamma\cos\psi + W_y \tag{9.5}$$

$$\dot{h} = v\sin\gamma + W_h \tag{9.6}$$

$$\dot{m} = -\frac{T}{u_e} \tag{9.7}$$

174

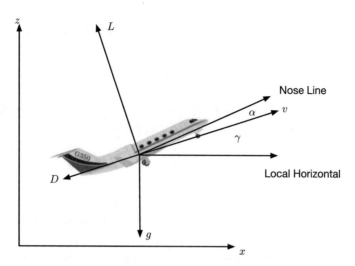

Figure 9.2: *Aircraft model showing lift, drag, and gravity.*

where v is the true airspeed, T is the thrust, L is the lift, g is the acceleration of gravity, γ is the air-relative flight path angle, ψ is the air-relative heading (measured clockwise from North), ϕ is the bank angle, x and y are the East and North positions, respectively, and h is the altitude. The mass is the total of dry mass and fuel mass. The terms $\{f_v, f_\gamma, f_\psi\}$ represent additional forces due to modeling uncertainty, and the terms $\{W_x, W_y, W_h\}$ are wind speed components. If the vertical wind speed is zero, then $\gamma = 0$ produces level flight. α, ϕ, and T are the controls. Figure 9.2 shows the longitudinal symbols for the aircraft. γ is the angle between the velocity vector and local horizontal. α is the angle of attack which is between the nose of the aircraft and the velocity vector. The wings may be oriented, or have airfoils, that give lift at zero angles of attack. Drag is opposite velocity and lift is perpendicular to drag. Lift must balance gravity and any downward component of drag; otherwise, the aircraft will descend.

We are using a very simple aerodynamic model. The lift coefficient is defined as

$$c_L = c_{L_\alpha}\alpha \tag{9.8}$$

The lift coefficient is a nonlinear function of the angle of attack. It has a maximum angle of attack above which the wing stalls and all lift is lost. For a flat plate, $c_{L_\alpha} = 2\pi$. The drag coefficient is

$$c_D = c_{D_0} + \frac{c_L^2}{\pi A_R \epsilon} \tag{9.9}$$

where A_R is the aspect ratio and ϵ is the Oswald efficiency factor which is typically from 0.8 to 0.95. The efficiency factor is how efficiently lift is coupled to drag. If it is less than one, it means that the lift produces more lift-induced drag than the ideal. The aspect ratio is the ratio of the wingspan (from the point nearest the fuselage to the tip) and the chord (the length from the front to the back of the wing).

The dynamic pressure, the pressure due to the motion of the aircraft, is

$$q = \frac{1}{2}\rho v^2 \tag{9.10}$$

where v is the speed and ρ is the atmospheric density. This is the pressure on your hand if you stick it out of the window of a moving car. The lift and drag forces are

$$L = qc_L s \tag{9.11}$$
$$D = qc_D s \tag{9.12}$$

where s is the wetted area. The wetted area is the surface of the aircraft that produces lift and drag. We make it the same for lift and drag, but in a real aircraft some parts of the aircraft cause drag (like the nose) but don't produce any lift. In essence, we assume the aircraft is all wing.

We create a right-hand-side function for the model. This will be called by the numerical integration function. The following has the dynamical model.

RHSPointMassAircraft.m

```
17
18  if( nargin < 1 )
19      xDot = DefaultDataStructure;
20      return
21  end
22
23  v        = x(1);
24  gamma    = x(2);
25  psi      = x(3);
26  h        = x(6);
27  cA       = cos(d.alpha);
28  sA       = sin(d.alpha);
29  cG       = cos(gamma);
30  sG       = sin(gamma);
31  cPsi     = cos(psi);
32  sPsi     = sin(psi);
33  cPhi     = cos(d.phi);
34  sPhi     = sin(d.phi);
35
36  mG       = d.m*d.g;
37  qS       = 0.5*d.s*Density( 0.001*h )*v^2;
38  cL       = d.cLAlpha*d.alpha;
39  cD       = d.cD0 + cL^2/(pi*d.aR*d.eps);
40  lift     = qS*cL;
41  drag     = qS*cD;
42  vDot     = (d.thrust*cA - drag - mG*sG)/d.m + d.f(1);
43  fN       = lift + d.thrust*sA;
44  gammaDot = (fN*cPhi - mG*cG + d.f(2))/(d.m*v);
45  psiDot   = (fN*sPhi - d.f(3))/(d.m*v*cG);
46  xDot     = [vDot;gammaDot;psiDot;v*cG*sPsi;v*cG*cPsi;v*sG];
```

The default data structure is defined in the subfunction, `DefaultDataStructure`. The data structure includes both constant parameters and control inputs.

```
50
51   d = struct('cD0',0.01,'aR',2.67,'eps',0.95,'cLAlpha',2*pi,'s',64.52,...
52             'g',9.806,'alpha',0,'phi',0,'thrust',0,'m',19368.00,...
53             'f',zeros(3,1),'W',zeros(3,1));
```

We use a modified exponential atmosphere for the density:

```
56   function rho = Density( h )
57
58   rho = 1.225*exp(-0.0817*h^1.15);
```

We want to maintain a force balance so that the speed of the aircraft is constant and the aircraft does not change its flight path angle. For example, in level flight the aircraft would not ascend or descend. We need to control the aircraft in level flight so that the velocity stays constant and $\gamma = 0$ for any ϕ. The relevant equations are

$$0 \;=\; T\cos\alpha - D \tag{9.13}$$

$$0 \;=\; (L + T\sin\alpha)\cos\phi - mg \tag{9.14}$$

We need to find T and α given ϕ.

A simple way is to use `fminsearch`. It will call `RHSPointMassAircraft` and numerically find controls that, for a given ψ, h and v have zero time derivatives. The following code finds the equilibrium angle of attack and thrust. RHS is called by `fminsearch`. It returns a scalar cost that is a quadratic of the acceleration (time derivative of velocity) and a derivative of the flight path angle. Our initial guess is a value of thrust that balances the drag. Even with an angle of attack guess of zero, it converges with the default set of parameters `opt = optimset('fminsearch')`.

EquilibriumControls.m

```
14   function d = EquilibriumControls( x, d )
15
16   if( nargin < 1 )
17     Demo
18       return
19   end
20
21   [~,~,drag]   = RHSPointMassAircraft( 0, x, d );
22   u0           = [drag;0];
23   opt          = optimset('fminsearch');
24   u            = fminsearch( @RHS, u0, opt, x, d );
```

```
25   d.thrust     = u(1);
26   d.alpha      = u(2);
27
28   %% EquilibriumControls>RHS
29   function c = RHS( u, x, d )
30
31   d.thrust     = u(1);
32   d.alpha      = u(2);
33   xDot         = RHSPointMassAircraft( 0, x, d );
34   c            = xDot(1)^2 + xDot(2)^2;
```

The demo is for a Gulfstream 350 flying at 250 m/s and 10 km altitude.

```
37   function Demo
38
39   d     = RHSPointMassAircraft;
40   d.phi = 0.4;
41   x     = [250;0;0.02;0;0;10000];
42   d     = EquilibriumControls( x, d );
43   r     = x(1)^2/(d.g*tan(d.phi));
44
45   fprintf('Thrust          %8.2f N\n',d.thrust);
46   fprintf('Altitude        %8.2f km\n',x(6)/1000);
47   fprintf('Angle of attack %8.2f deg\n',d.alpha*180/pi);
48   fprintf('Bank angle      %8.2f deg\n',d.phi*180/pi);
49   fprintf('Turn radius     %8.2f km\n',r/1000);
```

The results of the demo, in the following, are quite reasonable.

```
>> EquilibriumControls
Thrust          7614.63 N
Altitude          10.00 km
The angle of attack was 2.41 deg
Bank angle        22.92 deg
Turn radius       15.08 km
```

With these values, the plane will turn without changing altitude or airspeed. We simulate the Gulfstream in the script `AircraftSim.m`. The first part runs our equilibrium computation demo.

AircraftSim.m

```
1   %% Script to simulate a Gulfstream 350 in a banked turn
2
3   n   = 500;
4   dT  = 1;
5   rTD = 180/pi;
```

```
7
8    %% Start by finding the equilibrium controls
9    d       = RHSPointMassAircraft;
10   d.phi   = 0.4;
11   x       = [250;0;0.02;0;0;10000];
12   d       = EquilibriumControls( x, d );
13   r       = x(1)^2/(d.g*tan(d.phi));
14
15   fprintf('Thrust          %8.2f N\n',d.thrust);
16   fprintf('Altitude        %8.2f km\n',x(6)/1000);
17   fprintf('Angle of attack %8.2f deg\n',d.alpha*180/pi);
18   fprintf('Bank angle      %8.2f deg\n',d.phi*180/pi);
19   fprintf('Turn radius     %8.2f km\n',r/1000);
```

The next part does the simulation. It breaks the loop if the aircraft altitude is less than zero, that is, it crashes. We call `RHSPointMassAircraft` once to get the lift and drag value for plotting. It is then called by `RungeKutta` to do the numerical integration. @ denotes a pointer to the function.

```
21   %% Simulation
22   xPlot = zeros(length(x)+5,n);
23
24   for k = 1:n
25
26       % Get lift and drag for plotting
27       [~,L,D]    = RHSPointMassAircraft( 0, x, d );
28
29       % Plot storage
30       xPlot(:,k)  = [x;L;D;d.alpha*rTD;d.thrust;d.phi*rTD];
31
32       % Integrate
33       x           = RungeKutta( @RHSPointMassAircraft, 0, x, dT, d );
34
35       % A crash
36       if( x(6) <= 0 )
37           break;
38       end
39   end
```

The remainder produces three plots. The first plot is the states that are numerically integrated. The next gives the controls, lift, and drag. The final plot shows the planar trajectory. We do unit conversions since degrees and kilometers are a bit clearer.

```
41  %% Plot the results
42  xPlot          = xPlot(:,1:k);
43  xPlot(2,:)     = xPlot(2,:)*rTD;
44  xPlot(4:6,:)   = xPlot(4:6,:)/1000;
45  yL             = {'v (m/s)' '\gamma (deg)' '\psi (deg)' 'x_e (km)' 'y_n
         (km)'...
46                      'h (km)' 'L (N)' 'D (N)' '\alpha (deg)' 'T (N)' '\phi
                        (deg)'};
47  [t,tL]         = TimeLabel(dT*(0:(k-1)));
48
49  PlotSet( t, xPlot(1:6,:), 'x label', tL, 'y label', yL(1:6),...
50    'figure title', 'Aircraft State', 'ylim',{[249 251] [1.45
        1.46],[],[],[],[]} );
51  PlotSet( t, xPlot(7:11,:), 'x label', tL, 'y label', yL(7:11),...
52    'figure title', 'Aircraft Lift, Drag and Controls', 'ylim',{[2e5 2.1
        e5] [1.45e6 1.46e6],[],[],[],[]} );
```

As you can see in Figure 9.3, the radius of the turn is 15 km as expected. The drag and lift remain constant. In practice, we would have a velocity and flight path angle control system to handle disturbances or parameter variations. For our deep learning example, we just use the ideal dynamics. Figure 9.4 shows the simulation outputs.

Figure 9.3 will provide a nice trajectory for our deep learning examples. You can change the aircraft simulation to produce other trajectories.

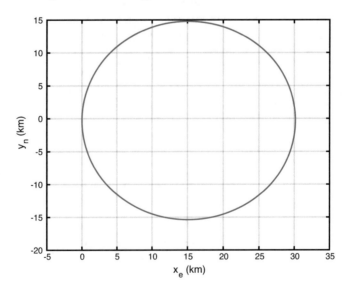

Figure 9.3: *Aircraft trajectory.*

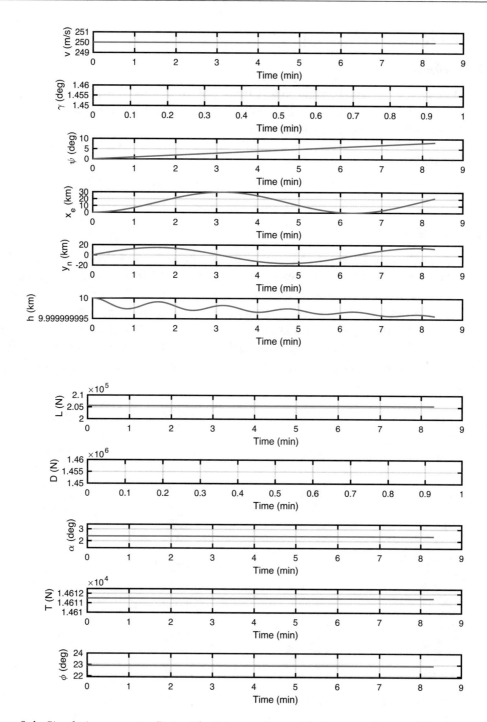

Figure 9.4: *Simulation outputs. States (the integrated quantities) are on the top. Lift, drag, and the controls ϕ, α, and T are on the bottom.*

9.3   Generating Terrain

9.3.1 Problem

We want to create an artificial terrain model from a set of terrain "tiles." A tile is a segment of terrain from a bigger picture, much like bathroom tiles make up a bathroom wall, unless, of course, you have the modern fiberglass shower.

9.3.2 Solution

Find images of terrain and tile them together. There are many sources of terrain tiles. Google Earth is one.

9.3.3 How It Works

We start by compiling a database of terrain tiles. We have them in the folder terrain in our MATLAB package. A segment of the terrain folder is shown in Figure 9.5. This is just one way to get terrain tiles. There are online sources for downloading tiles. Also, many flight simulator games have extensive terrain libraries. The name of the folder is latitude longitude. For example, -10-10 is −10 degrees latitude and −10 degrees longitude. Our database only extends to ± 60 degrees latitude. The first block creates a list of the folders in terrain. An important thing with this code is that your script needs to be in the correct directory. We don't do any fancy directory searching.

Figure 9.5: *A segment of the terrain folder.*

CreateTerrain.m

```
15  function CreateTerrain( lat, lon, scale )
16
17  % Demo
18  if( nargin < 1 )
19     Demo;
20     return
21  end
22
23  d              = dir('terrain');
24  latA           = zeros(1,468);
25  lonA           = zeros(1,468);
26  folderName  = cell(1,468);
27  for k = 1:468
28     q              = d(k).name;
29     folderName{k}  = q;
30     if( q(2) == '0' )
31        latA(k) = str2double(q(1:2));
32        lonA(k) = str2double(q(3:end));
33     else
34        latA(k) = str2double(q(1:3));
35        lonA(k) = str2double(q(4:end));
36     end
37  end
```

The next code block finds the indices for the desired tiles.

```
39  % Center lower left corner is start
40  latF    = floor(lat);
41  lonF    = floor(lon);
42  latI    = zeros(1,9);
43  lonI    = zeros(1,9);
44  lon0  = lonF - 10;
45  latJK = latF - 10;
46  lonJK = lon0;
47  i       = 1;
48  for j = 1:3
49     for k = 1:3
50        lonI(i) = lonJK;
51        latI(i) = latJK;
52        lonJK    = lonJK + 10;
53        i         = i + 1;
54     end
55     lonJK = lon0;
56     latJK = latJK + 10;
57  end
58
59  fldr = zeros(1,9);
```

```
60   for k = 1:9
61      j        = find(latI(k)==latA);
62      i        = lonI(k)==lonA(j);
63      fldr(k) = j(i);
64   end
```

The following code creates the filenames based on our latitudes and longitudes. We just create correctly formatted strings. This shows one way to create strings. Notice we use %d to create integers. It automatically makes them the right length. We need to check for positive and negative so that the + and - signs are correct.

```
66   % Generate the file names
67   imageSet  = cell(1,9);
68   for k = 1:9
69     j = fldr(k);
70     if( latA(j) >= 0  )
71       if( lonA(j) >= 0 )
72         imageSet{k} = sprintf('grid10x10+%d+%d',latA(j)*100,lonA(j)*100);
73       else
74         imageSet{k} = sprintf('grid10x10+%d-%d',latA(j)*100,lonA(j)*100);
75       end
76     else
77       if( lonA(j) >= 0 )
78         imageSet{k} = sprintf('grid10x10-%d+%d',latA(j)*100,lonA(j)*100);
79       else
80         imageSet{k} = sprintf('grid10x10-%d-%d',latA(j)*100,lonA(j)*100);
81       end
82     end
83   end
```

The next block reads in the image, flips it upside down, and scales the image. The images happen to be north down and south up. We first change directory to be in terrain and then cd to go into each folder. cd .. changes directories back into terrain.

```
85   % Assuming we are one directory above
86   cd terrain
87
88   im    = cell(1,9);
89   for k = 1:9
90           j = fldr(k);
91     cd(folderName{j})
92           im{k} = ScaleImage(flipud(imread([imageSet{k},'.jpg'])),scale);
93     cd ..
94   end
```

The next block of code calls image to draw each image in the correct spot on the 3 by 3 tiled map.

```
96   del     = size(im{1},1);
97   lX      = 3*del;
98
99   % Draw the images
100  x       = 0;
101  y       = 0;
102  for k = 1:9
103    image('xdata',[x;x+del],'ydata',[y;y+del],'cdata', im{k} );
104    hold on
105    x = x + del;
106    if ( x == lX )
107      x = 0;
108      y = y + del;
109    end
110  end
111  axis off
112  axis image
113
114  cd ..
```

The subfunction ScaleImage scales the image by doing a mean of the pixels that are scaled down to one pixel. At the very end, we cd .., putting us into the original directory.

```
116  %% CreateTerrain>ScaleImage
117  function s2 = ScaleImage( s1, q )
118
119  n = 2^q;
120
121  [mR,~,mD] = size(s1);
122
123  m = mR/n;
124
125  s2 = zeros(m,m,mD,'uint8');
126
127  for i = 1:mD
128    for j = 1:m
129      r = (j-1)*n+1:j*n;
130      for k = 1:m
131        c         = (k-1)*n+1:k*n;
132        s2(j,k,i) = mean(mean(s1(r,c,i)));
133      end
134    end
135  end
```

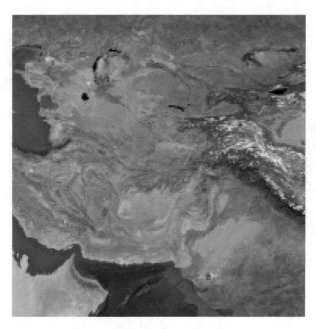

Figure 9.6: *Terrain tiled image of the Middle East.*

The demo picks a latitude and longitude in the Middle East. The result is that the 3 by 3 tiled image is shown in Figure 9.6. We can't use this image for the neural net because of its very low resolution.

```
137  %% CreateTerrain>Demo
138  function Demo
139
140  NewFigure('EarthSegment');
141  CreateTerrain( 30,60,1 )
```

9.4 Close-Up Terrain

9.4.1 Problem

We want higher-resolution terrain.

9.4.2 Solution

Specialize the terrain code to produce a small segment of higher-resolution terrain suitable for experiments with a commercial drone.

9.4.3 How It Works

The preceding terrain code would work well for an orbiting satellite, but not so well for a drone. Per FAA regulations, the maximum altitude for small unmanned aircraft is 400 feet or about 122 meters. A satellite in a low Earth orbit (LEO) typically has an altitude of 300–500 km. Thus, drones are typically about 2500–4000 times closer to the surface than a satellite! We take the code and specialize it to read in just four images. It is much simpler than `CreateTerrain` and is less flexible. If you want to change it, you will need to change the code in the file.

CreateTerrainClose.m

```matlab
8   function CreateTerrainClose
9
10  % Generate the file names
11  imageSet  = {'grid1x1+3400-11800','grid1x1+3400-11900',...
12               'grid1x1+3500-11800','grid1x1+3500-11900'};
13  p = [2 1 4 3];
14
15  % Assuming we are one directory above
16  cd terrainclose
17
18  im = cell(1,4);
19  for k = 1:4
20          im{k} = flipud(imread([imageSet{k},'.jpg']));
21  end
22
23  del = size(im{1},1);
24
25  % Draw the images
26  x       = 0;
27  y       = 0;
28  i       = 0;
29  for k = 1:2
30    for j = 1:2
31      i = i + 1;
32      image('xdata',[x;x+del],'ydata',[y;y+del],'cdata', im{p(i)} );
33      hold on
34      x = x + del;
35    end
36    x = 0;
37    y = y + del;
38  end
39  axis off
40  axis image
41
42  cd ..
```

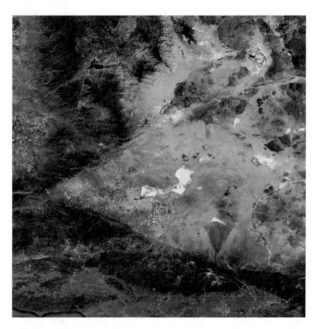

Figure 9.7: *Close-up terrain.*

We don't have any options for scaling. This runs the function:

```
>> NewFigure('EarthSegmentClose');
>> CreateTerrainClose
```

Figure 9.7 shows the terrain. It is 2 degrees by 2 degrees.

9.5 Building the Camera Model

9.5.1 Problem

We want to build a camera model for our deep learning system. We want a model that emulates the function of a drone-mounted camera. Ultimately, we will use this camera model as part of a terrain-based navigation system, and we'll apply deep learning techniques to do terrain navigation.

9.5.2 Solution

We will model a pinhole camera and create a high-altitude aircraft. A pinhole camera is a lowest-order approximation to a real optical system. We'll then build the simulation and demonstrate the camera.

9.5.3 How It Works

We've already created an aircraft simulation in Recipe 9.2. The addition will be the terrain model and the camera model. A pinhole camera is shown in Figure 9.8. A pinhole camera has an infinite depth of field, and the images are rectilinear.

A point $P(x, y, z)$ is mapped to the imaging plane by the relationships

$$u = \frac{fx}{h} \tag{9.15}$$

$$v = \frac{fy}{h} \tag{9.16}$$

where u and v are coordinates in the focal plane, f is the focal length, and h is the distance from the pinhole to the point along the axis normal to the focal plane. This assumes that the z-axis of the coordinate frame x, y, z is aligned with the boresight of the camera. The angle that is seen by the imaging chip is

$$\theta = \tan^{-1}\left(\frac{w}{2f}\right) \tag{9.17}$$

where f is the focal length. The shorter the focal length, the larger the image. The pinhole camera does not have any depth of field, but that is unimportant for this far-field imaging problem. The field of view of a pinhole camera is limited only by the sensing element. Most cameras have lenses, and the images are not perfect across the imaging array. This presents practical problems that need to be solved in real machine vision systems.

We want our camera to see 16 pixels by 16 pixels from the terrain image in Figure 9.7. We will assume a flight altitude of 10 km. Figure 9.9 gives the dimensions.

We are not simulating a particular camera. Instead, our camera model is producing 16 by 16 pixel maps given an input of a position. The output is a data structure with the x and y

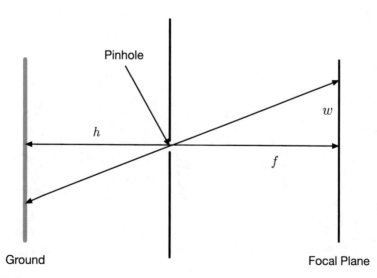

Figure 9.8: *Pinhole camera.*

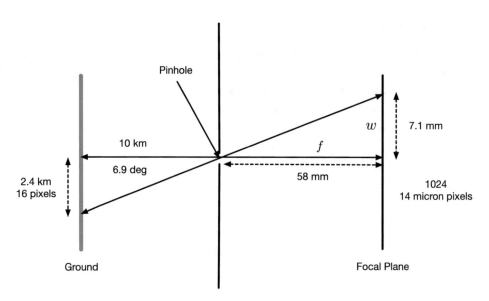

Figure 9.9: *Pinhole camera with dimensions.*

coordinates and an image. If no inputs are given, it will create a tiled map of the image. We scaled the image in the GraphicConverter app (Lemke Software GMBH) so that it is exactly

```
672    672    3
```

and saved it in the file `TerrainClose.jpg`. The numbers are x pixels, y pixels, and three layers for red, green, and blue. The third index is for the red, blue, and green matrices. This is a three-dimensional matrix, typical for color images.

The code is shown as follows. We convert everything to pixels, get the image using `[~,~,i] = getimage(h)`, and get the segment.

The first part of the code is to provide defaults for the user.

TerrainCamera.m

```
15  function d = TerrainCamera( r, h, nBits, w, nP )
16
17  % Demo
18  if( nargin < 1 )
19      Demo;
20      return
21  end
22
23  if( nargin < 3 )
24      nBits = [];
25  end
26
```

```
27   if( nargin < 4 )
28     w = [];
29   end
30
31   if( nargin < 5 )
32     nP = 64;
33   end
34
35   if( isempty(w) )
36     w = 4000;
37   end
38
39   if( isempty(nBits) )
40     nBits = 16;
41   end
```

The next part computes the pixels.

TerrainCamera.m

```
43   dW = w/nP;
44
45   x0 = -w/2 + (nBits/2)*dW;
46   y0 =  w/2 - (nBits/2)*dW;
47   k  = floor((r(1) - x0)/dW) + 1;
48   j  = floor((y0 - r(2))/dW) + 1;
```

The remainder displays the image.

TerrainCamera.m

```
51   kR = k:(k-1 + nBits);
52   kJ = j:(j-1 + nBits);
53
54   [~,~,i] = getimage(h);
55
56   d.p     = i(kR,kJ,:);
```

The demo draws the source image and then the camera image. Both are shown in Figure 9.10.

```
58
59   if( nargout < 1 )
60     image(d.p)
61     axis off
62     axis image
63     clear p
64   end
```

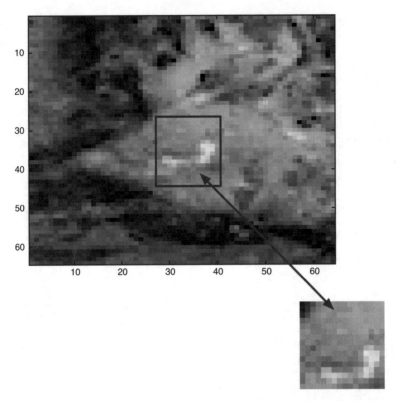

Figure 9.10: *Terrain camera source image and camera view. The camera view is 16 × 16 pixels.*

```
65
66   %% CreateTerrain>Demo
67   function Demo
68
69   h = NewFigure('Earth Segment');
70   i = imread('TerrainClose64.jpg');
71   image(i);
72   grid
```

The terrain image from the camera is blurry because it has so few pixels.

9.6 Plotting the Trajectory

9.6.1 Problem

We want to plot our trajectory over an image.

9.6.2 Solution

Create a function to draw the image and plot the trajectory on top.

9.6.3 How It Works

We write a function that reads in an image and plots the trajectory on top. We scale the image using $\texttt{image}$. The x-dimension is set and the y-dimension is scaled to match.

PlotXYTrajectory.m

```
1   %% PLOTXYTRAJECTORY Draw an xy trajectory over an image
2   % Can plot multiple sets of data. Type PlotXYTrajectory for a demo.
3   %% Input
4   % x         (:,:) X coordinates (m)
5   % y         (:,:) Y coordinates (m)
6   % i         (n,m) Image
7   % w         (1,1) x dimension of the image
8   % xScale    (1,1) Scale of x dimension
9   % name      (1,:) Figure name
10
11  function PlotXYTrajectory( x, y, i, xScale, name )
12
13  if( nargin < 1 )
14    Demo
15    return
16  end
17
18  s    = size(i);
19  xI   = [-xScale xScale];
20  yI   = [-xScale xScale]*s(2)/s(1);
21
22  NewFigure(name);
23  image(xI,yI,flipud(i));
24  hold on
25  n = size(x,1);
26  for k = 1:n
27    plot(x(k,:),y(k,:),'linewidth',2)
28  end
29  set(gca,'xlim',xI,'ylim',yI);
30  grid on
31  axis image
32  xlabel('x (m)')
33  ylabel('y (m)')
```

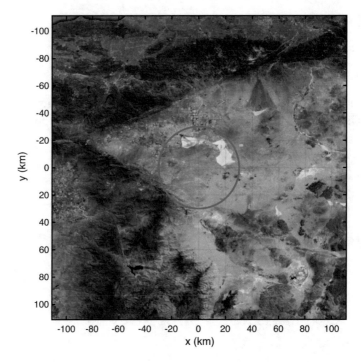

Figure 9.11: *Trajectory plot.*

The demo draws a circle over our terrain image. This is shown in Figure 9.11.

```
36   %% PlotXYTrajectory>Demo
37   function Demo
38
39   i = imread('TerrainClose.jpg');
40   a = linspace(0,2*pi);
41   x = [30*cos(a);35*cos(a)];
42   y = [30*sin(a);35*sin(a)];
43   PlotXYTrajectory( x, y, i, 111, 'Trajectory' )
```

While the deep learning system will analyze all of the pixels in the image, it is interesting to see how the mean values of the pixel colors vary across an image. This is shown in Figure 9.12. The x-axis is the image number, going by rows of constant y. As can be seen, there is considerable variation even in nearby images. This indicates that there is sufficient information in each image for our deep learning system to be able to find locations. It also shows that it might be possible just to use mean values to identify the location. Remember that each image varies from the previous by only 16 pixels.

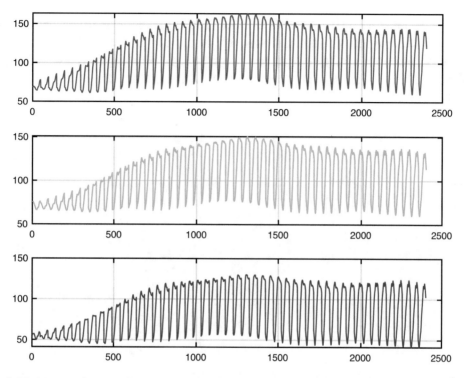

Figure 9.12: *Mean red, green, blue values for the images.*

9.7 Creating the Training Images

9.7.1 Problem

We want to create training images for our terrain model.

9.7.2 Solution

We build a script to read in the 64 by 64 bit image and create training images.

9.7.3 How It Works

We first create a 64-bit version of our terrain, using any image processing app. We've already done that, and it is saved as `TerrainClose64.jpg`. The following script reads in the image and generates training images by displacing the index one pixel at a time. We save the images in the folder `TerrainImages`. We also create labels. Each image is a different label. For each terrain snippet, we create nN copies with noise. Thus, there will be nN images with the label 1. We add noise with the code

```
uint8(floor(sig*rand(nBits,nBits,3)))
```

since the noise must be `uint8` like the image. You'll get an error if you don't convert to `uint8`. You can also select different strides, that is, moving the images more than one pixel. The first code sets up the image processing. We choose 16-bit images because (after the next step of training) there is enough information in each image to classify each one. We tried 8 bits but it didn't converge.

CreateTerrainImages.m

```
13  im    = flipud(imread('TerrainClose64.jpg')); % Read in the image
14  wIm   = 4000; % m
15  nBits = 16;
16  dN    = 1; % The delta bits is 1
17  nBM1  = nBits-1;
18  [n,m] = size(im); % Size of the image
19  nI    = (n-nBits)/dN + 1; % The number of images down one side
20  nN    = 10;      % How many copies of each image we want
21  sig   = 3;       % Set to > 0 to add noise to the images
22  dW    = wIm/64; % Delta position for each image (m)
23  x0    = -wIm/2+(nBits/2)*dW;   % Starting location in the upper left
        corner
24  y0    =  wIm/2-(nBits/2)*dW;   % Starting location in the upper left
        corner
```

This line is very important. It makes sure the names correspond to distinct images. We will make copies of each image for training purposes.

CreateTerrainImages.m

```
26  % Make an image serial number so they remain in order in the
        imageDatastore
27  kAdd = 10^ceil(log10(nI*nI*nN));
```

We do some directory manipulations here.

CreateTerrainImages.m

```
29  % Set up the directory
30  if ~exist('TerrainImages','dir')
31    warning('Are you in the right folder? No TerrainImages')
32    [success,msg] = mkdir('./','TerrainImages')
33  end
34  cd TerrainImages
35  delete *.jpg % Starting from scratch so delete existing images
```

The image splitting is done in this code. We add noise, if desired.

CreateTerrainImages.m

```
37   kCheck = randperm(nI-1,2);
38
39   i        = 1;
40   l        = 1;
41   t        = zeros(1,nI*nI*nN); % The label for each image
42   x        = x0; % Initial location
43   y        = y0; % Initial location
44   r        = zeros(2,nI*nI); % The x and y coordinates of each image
45   id       = zeros(1,nI*nI);
46   iMI      = zeros(1,nI*nI);
47   rgbs     = [];
48   hW       = waitbar(0,'Processing Terrain Images');
49
50   for k = 1:nI
51     waitbar(k/nI,hW);
52     kR = dN*(k-1)+1:dN*(k-1) + nBits;
53     for j = 1:nI
54       kJ             = dN*(j-1)+1:dN*(j-1) + nBits;
55       thisImg        = im(kR,kJ,:);
56       rgbs(end+1,:) = [mean(mean(thisImg(:,:,1))) mean(mean(thisImg
                (:,:,2))) mean(mean(thisImg(:,:,3)))];
57       for p = 1:nN
58         s            = im(kR,kJ,:) + uint8(floor(sig*rand(nBits,nBits,3)));
59         q            = s > 256;
60         s(q)         = 256;
61         q            = s < 0;
62         s(q)         = 0;
63         imwrite(s,sprintf('TerrainImage%d.jpg',i+kAdd));
64         t(i)         = 1;
65         i            = i + 1;
66       end   % number of images at each location
67       if (k==kCheck(1) && j==kCheck(2))
68         imgCheck = thisImg;
69         rCheck = [x;y];
70       end
71       r(:,1)         = [x;y];
```

Figure 9.13 shows that the images cover the area. We also verified that the sum of R, G, and B was different for each image. This indicates that there is enough information for the machine learning algorithm.

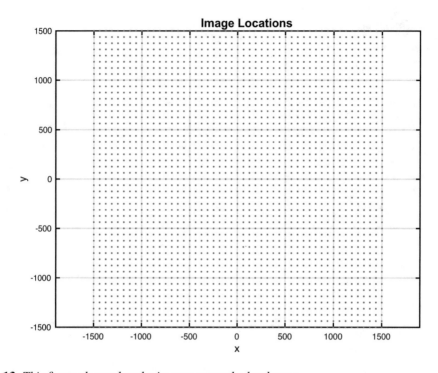

Figure 9.13: *This figure shows that the images cover the landscape.*

9.8 Training and Testing

9.8.1 Problem

We want to create and test a convolutional neural network. The neural net will be trained to associate images with an x and y location.

9.8.2 Solution

We create and test a convolutional neural network in `TerrainNeuralNet.m`. This will be trained on the images created earlier and will be able to return the x and y coordinates. Convolutional neural networks are widely used for image identification.

9.8.3 How It Works

This example is much like the one in Chapter 3. The difference is that each image is a separate category. This is like face identification where each category is a different person.

TerrainNeuralNet.m

```
1   %% Script implementing the terrain neural net
2   % You must have created the images in TerrainImages with
      CreateTerrainImages
3   % before running this script.
4
5   %% Get the images
6   cd TerrainImages
7   label = load('Label');
8   cd ..
9
10  t           = categorical(label.t);
11  nClasses    = max(label.t);
12  imds        = imageDatastore('TerrainImages','labels',t);
13  labelCount = countEachLabel(imds);
14
15  % Display a few snapshots
16  NewFigure('Terrain Snapshots');
17  n = 4;
18  m = 5;
19  ks = sort(randi(length(label.t),1,n*m)); % random selection
20  for i = 1:n*m
21          subplot(n,m,i);
22          imshow(imds.Files{ks(i)});
23      title(sprintf('Image %d: %d',ks(i),label.t(ks(i))))
24  end
25
26  % We need the size of the images for the input layer
27  img = readimage(imds,1);
28
29  % Split into training and testing sets
30  fracTraining = 0.8;
31  [imdsTrain,imdsTest] = splitEachLabel(imds,fracTraining,'randomized');
32
33  %% Training
34  % This gives the structure of the convolutional neural net
35  layers = [
36      imageInputLayer(size(img))
37
38      convolution2dLayer(3,8,'Padding','same')
39      batchNormalizationLayer
40      reluLayer
41
42      maxPooling2dLayer(2,'Stride',2)     % Pool size and stride size
43
44      convolution2dLayer(3,32,'Padding','same')
45      batchNormalizationLayer
46      reluLayer
47
48      maxPooling2dLayer(2,'Stride',2)
49
```

```
50      fullyConnectedLayer(nClasses)
51      softmaxLayer
52      classificationLayer
53          ];
54  disp(layers)
55
56  options = trainingOptions('sgdm', ...
57      'InitialLearnRate',0.01, ...
58      'MaxEpochs',6, ...
59      'MiniBatchSize',100,...
60      'ValidationData',imdsTest, ...
61      'ValidationFrequency',10, ...
62      'ValidationPatience',inf,...
63      'Shuffle','every-epoch', ...
64      'Verbose',false, ...
65      'Plots','training-progress');
66  disp(options)
67  fprintf('Fraction for training %8.2f%%\n',fracTraining*100);
69
70  terrainNet = trainNetwork(imdsTrain,layers,options);
71
72  %% Test the neural net
73  predLabels  = classify(terrainNet,imdsTest);
74  testLabels  = imdsTest.Labels;
75
76  accuracy = sum(predLabels == testLabels)/numel(testLabels);
77  fprintf('Accuracy is %8.2f%%\n',accuracy*100)
78
79  save('TerrainNet','terrainNet')
```

We have an image layer to read in each image. We next convolve them with filters. The weights of the filters are determined during the learning. We normalize the outputs and pass through the ReLU activation function. Pooling compresses the data. Padding sets the output size equal to the input size. As seen by the layers printout, no padding is needed since the images are all the same size. The first layer has eight 3 by 3 pixel filters. The second layer has 32 3 by 3 pixel filters. The final set of layers is used to classify the images. As noted in the previous section, each image has a unique "class" which is associated with its location. We use a constant learning rate. The batch size is smaller than the default.

Figure 9.14 shows some of the images. Figure 9.15 shows the training window. It can categorize the images after seven epochs. The difference between the two adjacent images is only 16 pixels. It isn't a lot of data, but the neural net can categorize each image with 100% accuracy.

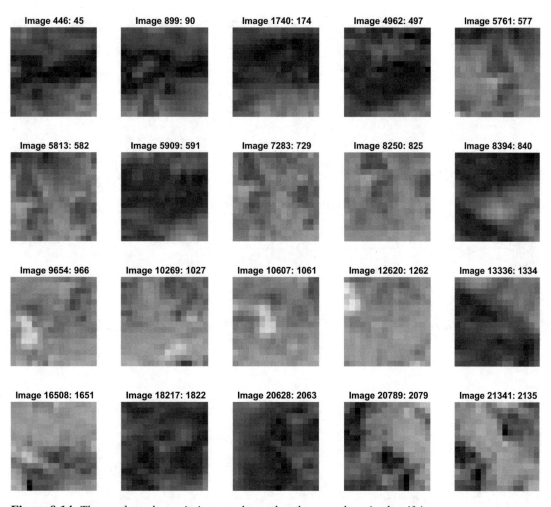

Figure 9.14: *These selected terrain images show what the neural net is classifying.*

In each epoch in Figure 9.15, it is processing all of the training data.

```
>> TerrainNeuralNet
   12x1 Layer array with layers:

      1   ''   Image Input           16x16x3 images with 'zerocenter'
           normalization
      2   ''   Convolution           8 3x3 convolutions with stride [1
           1] and padding 'same'
      3   ''   Batch Normalization   Batch normalization
      4   ''   ReLU                  ReLU
      5   ''   Max Pooling           2x2 max pooling with stride [2   2]
           and padding [0   0   0   0]
      6   ''   Convolution           32 3x3 convolutions with stride [1
           1] and padding 'same'
```

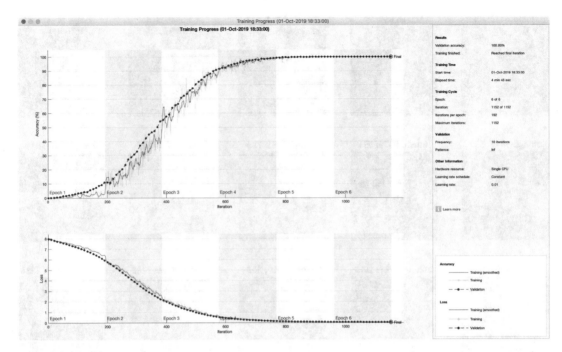

Figure 9.15: *Training window.*

```
       7    ''    Batch Normalization      Batch normalization
       8    ''    ReLU                     ReLU
       9    ''    Max Pooling              2x2 max pooling with stride [2  2]
            and padding [0  0  0  0]
      10    ''    Fully Connected          2401 fully connected layer
      11    ''    Softmax                  softmax
      12    ''    Classification Output    crossentropyex
TrainingOptionsSGDM with properties:

                   Momentum: 0.9000
           InitialLearnRate: 0.0100
  LearnRateScheduleSettings: [1x1 struct]
           L2Regularization: 1.0000e-04
     GradientThresholdMethod: 'l2norm'
          GradientThreshold: Inf
                  MaxEpochs: 6
              MiniBatchSize: 100
                    Verbose: 0
           VerboseFrequency: 50
             ValidationData: [1x1 matlab.io.datastore.ImageDatastore]
        ValidationFrequency: 10
         ValidationPatience: Inf
                    Shuffle: 'every-epoch'
             CheckpointPath: ''
       ExecutionEnvironment: 'auto'
                 WorkerLoad: []
```

202

```
             OutputFcn: []
                 Plots: 'training-progress'
        SequenceLength: 'longest'
   SequencePaddingValue: 0
    DispatchInBackground: 0
 Fraction for training    80.00%
 Accuracy is    100.00%
```

We get 100% accuracy. You can explore changing the number of layers and trying different activation functions. Training takes a few minutes.

9.9   Simulation

9.9.1 Problem

We want to test our deep learning algorithm using our terrain model.

9.9.2 Solution

We build a simulation using the trained neural net.

9.9.3 How It Works

We reproduce the simulation from the previous section and remove some unneeded output so that we can focus on the neural net. We read in the trained neural net.

AircraftNNSim.m

```
10  %% Load the neural net
11  nN    = load('TerrainNet');
12  rI    = load('Loc');
```

The neural net classifies the image obtained by the camera. We convert the category into an integer using `int32`. The `subplot` displays the image the neural net identifies as matching the camera image and the camera image. The simulation loop stops if your altitude, `x(6)`, is less than 1.

```
14  %% Start by finding the equilibrium controls
15  d     = RHSPointMassAircraft;
16  v     = 120;
17  d.phi = atan(v^2/(r*d.g));
18  x     = [v;0;0;-r;0;10000];
19  d     = EquilibriumControls( x, d );
20
21  %% Simulation
22  xPlot = zeros(length(x)+3,n);
23
24  % Put the image in a figure so that we can read it
25  h = NewFigure('Earth Segment');
26  i = flipud(imread('TerrainClose64.jpg'));
```

```
27  image(i);
28  axis image
29
30  NewFigure('Camera');
31
32  for k = 1:n
33
34    % Get the image for the neural net
35    im            = TerrainCamera( x(4:5), h, nBits );
36    subplot(1,2,1)
37    image(im.p)
38    axis image
39
40    % Run the neural net
41    l             = classify(nN.terrainNet,im.p);
42    subplot(1,2,2)
43    q = imread(sprintf('TerrainImages/TerrainImage%d.jpg',rI.iMI(l)));
44    image(q);
45    axis image
46
47    % Plot storage
48    i             = int32(l);
49    xPlot(:,k)    = [x;rI.r(:,i);i];
50
51    % Integrate
52    x             = RungeKutta( @RHSPointMassAircraft, 0, x, dT, d );
53
54    % A crash
55    if( x(6) <= 0 )
56      break;
57    end
58  end
```

Figure 9.16 shows the trajectory and the camera view. We simulate one full circle.

The identified terrain segment and the path, based on the neural network location, are shown in Figure 9.17. The neural net classifies the terrain it is seeing. The location of each image is read out and used to plot the trajectory.

The 2D trajectory is shown in Figure 9.18 for a circular path. We make sure we are in the regions where each image is a pixel change. On the edges, there is one image border. If we were in that region, the resolution would be below. The trajectory from the images is reasonably close to the actual trajectory. Better results would require higher resolution. In practice, the measured positions would be inputs to a Kalman filter [42] that modeled the aircraft dynamics, given earlier in this chapter. An input to the Kalman filter could be a 3-axis accelerometer (see Chapter 7). This would smooth the trajectory and improve accuracy.

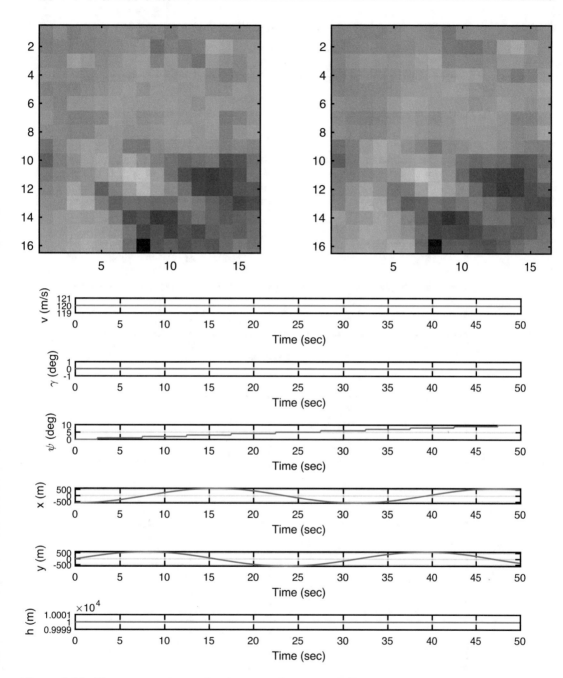

Figure 9.16: *The camera view and trajectory. This is one full circle. The two images are one pixel different.*

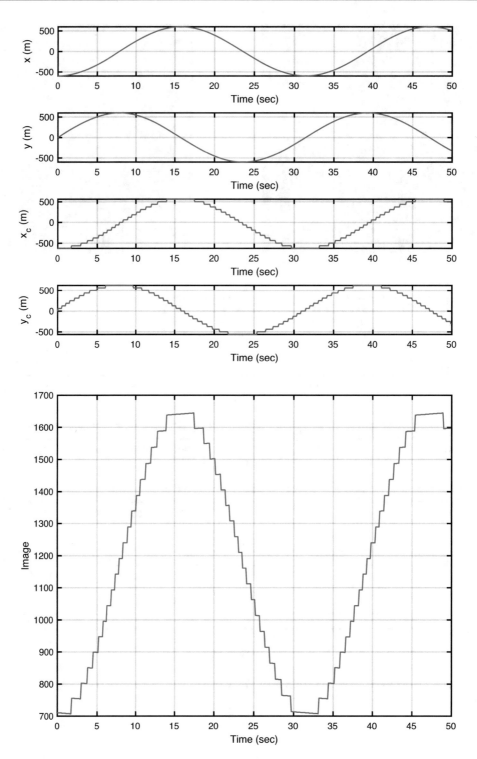

Figure 9.17: *The aircraft path and the identified terrain segments. "Image" in the bottom plot refers to the image index.*

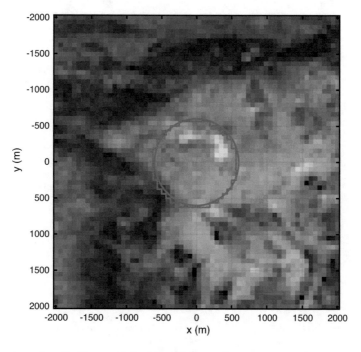

Figure 9.18: *The aircraft path, blue, and the identified terrain segments, red.*

This chapter showed how a neural network can be used to identify terrain for aircraft navigation. We simplified things by flying at a constant altitude, using a pinhole camera model with fixed image orientation, and ignoring clouds and other complications. We used a convolutional neural network to train the neural net with good results. As noted, higher-resolution images and a Kalman filter would produce a smoother trajectory.

CHAPTER 10

■ ■ ■

Stock Prediction

10.1 Introduction

The goal of a stock prediction algorithm is to recommend a portfolio of stocks that will maximize an investor's return. The investor has a finite amount of money and wants to create a portfolio to maximize her or his return on investment. The neural network in this chapter will predict the behavior of a portfolio of stocks given its history. This could then be used to select a portfolio of stocks with some idea of the future performance. The stock market model used in this chapter is based on Geometric Brownian Motion. Given that, we could do statistical analysis that would allow us to pick stocks. We'll show that a neural net, which does not have any knowledge of the model, can do as well in modeling the stocks.

10.2 Generating a Stock Market

10.2.1 Problem

We want to create an artificial stock market that replicates real stocks.

10.2.2 Solution

Implement Geometric Brownian Motion. This was invented by Paul Samuelson, Nobel Laureate [36].

10.2.3 How It Works

Paul Samuelson [11] created a stock model based on Geometric Brownian Motion. This approach produces realistic numbers and will not go negative. This is effectively a random walk in log space. The stochastic differential equation is

$$dS(t) = rSdt + \sigma SdW(t) \tag{10.1}$$

© Michael Paluszek, Stephanie Thomas, Eric Ham 2022
M. Paluszek et al., *Practical MATLAB Deep Learning*,
https://doi.org/10.1007/978-1-4842-7912-0_10

S is the stock price. $W(t)$ is a Brownian, random walk, process. t is the time, and dt is the time differential. r is the drift, and σ is the volatility of the stock market. Both range from zero to one. It could also be written in differential equation form as

$$\frac{dS}{dt} = \left(r + \sigma \frac{dW(t)}{dt} \right) S \tag{10.2}$$

The solution is

$$S(t) = S(0)e^{\left[(r - \frac{1}{2}\sigma^2)t + \sigma W(t) \right]} \tag{10.3}$$

The following shows the code used to generate the stock trends. We use cumsum to sum the random numbers for the random walk. We use a Gaussian or normal distribution produced by randn to create the random numbers. The function can create multiple stocks.

StockPrice.m

```
27  function [s, t] = StockPrice( s0, r, sigma, tEnd, nInt )
28
29  if( nargin < 1 )
30    Demo
31    return
32  end
33
34  delta   = tEnd/nInt;
35  sDelta  = sqrt(delta);
36  t       = linspace(0,tEnd,nInt+1);
37  m       = length(s0);
38  w       = [zeros(m,1) cumsum(sDelta.*randn(m,nInt))];
39  s       = zeros(1,nInt+1);
40  f       = r - 0.5*sigma.^2;
41  for k = 1:m
42    s(k,:) = s0(k)*exp(f(k)*t + sigma(k)*w(k,:));
43  end
```

The demo is based on the Wilshire 5000 statistics. It is an index of all US stocks. If you run it, you will get different values since the input is random.

```
51  %% StockPrice>Demo
52  function Demo
53
54  tEnd  = 5.75;
55  n     = 1448;
56  s0    = 8242.38;
57  r     = 0.1682262;
58  sigma = 0.1722922;
59  StockPrice( s0, r, sigma, tEnd, n );
60
61  %% Copyright
62  %    Copyright (c) 2019 Princeton Satellite Systems, Inc.
63  %    All rights reserved.
```

The results are shown in Figure 10.1. They look like a real stock. Changing the drift or volatility will change the overall shape. For example, if you set the volatility, $\sigma = 0$, you get the very nice stock shown with the red line. Increasing r makes the stock grow faster. This gives us the general rule that we want high r and low σ. See Figure 10.2.

Our model is based on two coefficients. We could make a stock-picking algorithm by just fitting stock price curves and computing r and σ. However, we want to see how well deep

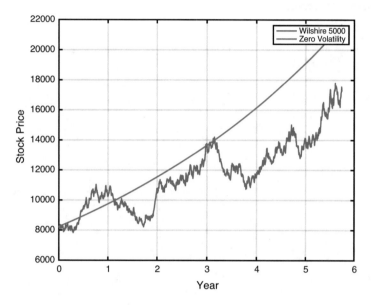

Figure 10.1: *A random stock based on statistics from the Wilshire 5000 vs. a stock with zero volatility. If you run the* StockPrice *function multiple times, you will get different results.*

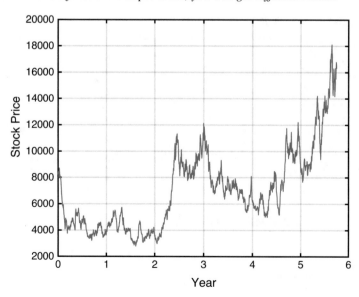

Figure 10.2: *A stock with high volatility and low drift such that* $r - \frac{1}{2}\sigma^2 < 1$. *In this case,* $r = 0.1$ *and* $\sigma = 0.6$.

211

learning does. Remember, this is a simple model of stock prices. Both σ and r could also be functions of time or random variables by themselves. Of course, there are other stock models too! The idea here is that deep learning creates its internal model without a need to be told about the model underlying the observed data.

The function `PlotStock.m` plots the stock price. Notice that we format the y tick labels ourselves to get rid of the exponential format that MATLAB would normally employ. `gca` returns the current axes handle.

PlotStock.m

```
12  function PlotStock(t,s,symb)
13
14  if( nargin < 1 )
15    Demo;
16    return;
17  end
18
19  m = size(s,1);
20
21  PlotSet(t,s,'x label','Year','y label','Stock Price','figure title',...
22      'Stocks','Plot Set',{1:m},'legend',{symb});
23
24  % Format the ticks
25  yT  = get(gca,'YTick');
26  yTL = cell(1,length(yT));
27  for k = 1:length(yT)
28        yTL{k} = sprintf('%5.0f',yT(k));
29  end
30  set(gca,'YTickLabel', yTL)
```

The built-in demo is the same as in `StockPrice`.

```
34  function Demo
35
36  tEnd  = 5.75;      % years
37  nInt  = 1448;      % intervals
38  s0    = 8242.38;   % initial price
39  r     = 0.1682262; % drift
40  sigma = 0.1722922;
41  [s,t] = StockPrice( s0, r, sigma, tEnd, nInt );
42  PlotStock(t,s,{})
```

10.3 Creating a Stock Market

10.3.1 Problem

We want to create a stock market.

10.3.2 Solution

Use the stock price function to create 100 stocks with randomly chosen parameters.

10.3.3 How It Works

We write a function that randomly picks stock starting prices, volatility, and drift. It also creates random three-letter stock names. We use a half-normal distribution for the stock prices. This code generates the random market. We limit the drift to between 0 and 0.5. This creates more stocks (for small markets) that go down.

StockMarket.m

```
19  function d = StockMarket(  nStocks, s0Mean, s0Sigma, tEnd, nInt )
20
21  if( nargin < 1 )
22    Demo
23    return
24  end
25
26  d.s0    = abs(s0Mean + s0Sigma*randn(1,nStocks));
27  d.r     = 0.5*rand(1,nStocks);
28  d.sigma = rand(1,nStocks);
29  s       = 'A':'Z';
30  for k = 1:nStocks
31    j           = randi(26,1,3);
32    d.symb(k,:) = s(j);
33  end
```

The following code plots all of the stocks in one plot. We create a legend and make the y labels integers (using `PlotStock`).

```
35  % Output
36  if( nargout < 1 )
37    s = StockPrice( d.s0, d.r, d.sigma, tEnd, nInt );
38    t       = linspace(0,tEnd,nInt+1);
39    PlotStock(t,s,d.symb);
40    clear d
41  end
```

The demo is shown as follows:

```
43  %% StockPrice>Demo
44  function Demo
45
46  nStocks = 15;     % number of stocks
47  s0Mean  = 8000;   % Mean stock preice
48  s0Sigma = 3000;   % Standard  dev of price
49  tEnd    = 5.75;   % years duration for market
50  nInt    = 1448;   % number of intervals
51  StockMarket( nStocks, s0Mean, s0Sigma, tEnd, nInt );
```

Two runs are shown in Figure 10.3.

A stock market with a hundred stocks is shown in Figure 10.4.

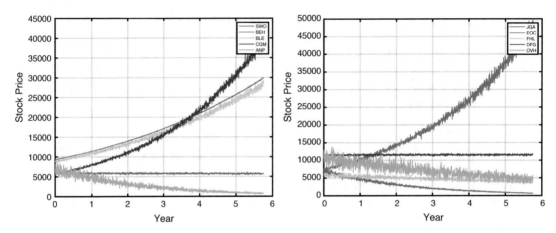

Figure 10.3: *Two runs of random five stock markets.*

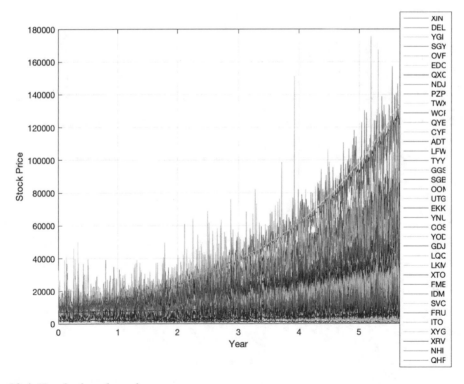

Figure 10.4: *Hundred stock market.*

10.4   Training and Testing

10.4.1 Problem

We want to build a deep learning system to predict the performance of a stock. This can be applied to the stock market created earlier to predict the performance of a portfolio.

10.4.2 Solution

The history of a stock is a time series. We will use a Long Short-Term Memory (LSTM) network to predict the future performance of the stock based on past data. Past performance is not necessarily indicative of future results. All investments carry some amount of risk. You are encouraged to consult with a certified financial planner before making any investment decisions. This utilizes the Deep Learning Toolbox's `lstmLayer` layer. We will use part of the time series to test the results.

10.4.3 How It Works

An LSTM layer learns long-term dependencies between time steps in a time series. It automatically deweights past data. LSTMs have replaced recursive neural nets (RNNs) in many applications.

The script `StockMarketNeuralNet.m` implements the neural network. The first part creates a market with a single stock. We set the random number generator to its default value, `rng('default')`, so that every time you run the script, you will get the same result. If you remove this line, you will get different results each time. The neural network training data is the time sequence, and the time sequence is shifted by a one-time step.

StockMarketNeuralNet.m

```
 1  %% Script using LSTM to predict stock prices
 2  %% See also:
 3  % lstmLayer, sequenceInputLayer, fullyConnectedLayer, regressionLayer,
 4  % trainingOptions, trainNetwork, predictAndUpdateState
 5
 6  % Rest the random number generator so we always get the same case
 7  rng('default')
 8
 9  layerSet = 'two lstm'; % 'lstm' 'bilstm' and 'two lstm' are available
10
11  %% Generate the stock market example
12  n      = 1448;
13  tEnd   = 5.75;
14  d      = StockMarket( 1, 8000, 3000, tEnd, n );
15  s      = StockPrice( d.s0, d.r, d.sigma, tEnd, n );
16  t      = linspace(0,tEnd,n+1);
17
18  PlotStock(t,s,d.symb);
```

The stock price is shown in Figure 10.5. We divide the outputs into training and testing data, using the first 80% of the data for training. We normalize the data which can produce better results when data has a large range. For simplicity in this example, we use the same data for validation and testing, although in a production system these should be different.

StockMarketNeuralNet.m

```
20   %% Divide into training and testing data
21   n            = length(s);
22   nTrain       = floor(0.8*n);
23   sTrain       = s(1:nTrain);
24   sTest        = s(nTrain+1:n);
25   sVal         = sTest;
26
27   % Normalize the training data
28   mu           = mean(sTrain);
29   sigma        = std(sTrain);
30
31   sTrainNorm   = (sTrain-mu)/sigma; % normalize the data to zero mean
32
33   % Normalize the test data
34   sTestNorm    = (sTest - mu) / sigma;
35   sTest        = sTestNorm(1:end-1);
```

The next part of the script sets up and trains the network. We use the "Adam" method [20]. Adam is a first-order gradient-based optimization of stochastic objective functions. It is computationally efficient and works well with problems with noisy or sparse gradients. See the reference for more details. We have a four-layer network including an LSTM layer.

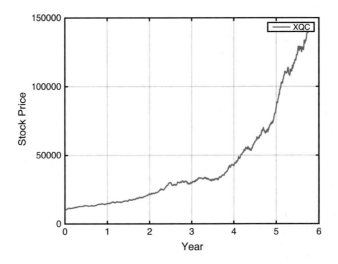

Figure 10.5: *A stock price.*

```
37   %% Set up and train the neural net
38
39   % We are training the LSTM using the previous step
40   xTrain     = sTrainNorm(1:end-1);
41   yTrain     = sTrainNorm(2:end);
42
43   % Validation data
44   muVal      = mean(sVal); % Must normalize over just this data
45   sigmaVal   = std(sVal);
46   sValNorm   = (sVal-muVal)/sigmaVal;
47
48   xVal = sValNorm(1:end-1);
49   yVal = sValNorm(2:end);
50
51   numFeatures   = 1;
52   numResponses  = 1;
53   numHiddenUnits = 200;
54
55   switch layerSet
56     case 'lstm'
57       layers = [sequenceInputLayer(numFeatures)
58                 lstmLayer(numHiddenUnits)
59                 fullyConnectedLayer(numResponses)
60                 regressionLayer];
61     case 'bilstm'
62       layers = [sequenceInputLayer(numFeatures)
63                 bilstmLayer(numHiddenUnits)
64                 fullyConnectedLayer(numResponses)
65                 regressionLayer];
66     case 'two lstm'
67       layers = [sequenceInputLayer(numFeatures)
68                 lstmLayer(numHiddenUnits)
69                 reluLayer
70                 lstmLayer(numHiddenUnits)
71                 fullyConnectedLayer(numResponses)
72                 regressionLayer];
73     otherwise
74       error('Only 3 sets of layers are available');
75   end
76
77   analyzeNetwork(layers);
78
79   options = trainingOptions('adam', ...
80       'MaxEpochs',300, ...
81       'ExecutionEnvironment','gpu',...
82       'GradientThreshold',1, ...
83       'InitialLearnRate',0.005, ...
84       'LearnRateSchedule','piecewise', ...
85       'LearnRateDropPeriod',125, ...
86       'LearnRateDropFactor',0.2, ...
87       'Shuffle','every-epoch', ...
```

```
88      'ValidationData',{xVal,yVal}, ...
89      'ValidationFrequency',5, ...
90      'Verbose',0, ...
91      'Plots','training-progress');
92
93  net = trainNetwork(xTrain,yTrain,layers,options);
```

The minimum number of layers when using the LSTM neural net is four, as shown in the following:

```
layers = [sequenceInputLayer(numFeatures)
          lstmLayer(numHiddenUnits)
          fullyConnectedLayer(numResponses)
          regressionLayer];
```

The layer structure is shown by `analyzeNetwork` in Figure 10.6. `analyzeNetwork` isn't too interesting for such a simple structure. It is more interesting when you have dozens or hundreds of layers. The script also provides the option to try a `bilstm` layer and two `lstm` layers:

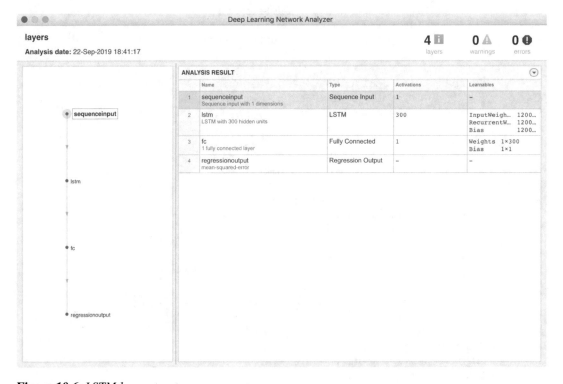

Figure 10.6: *LSTM layer structure.*

1. `sequenceInputLayer(inputSize)` defines a sequence input layer. `inputSize` is the size of the input sequence at each time step. In our problem, the sequence is just the last value in the time sequence, so `inputSize` is 1. You could have longer sequences.

2. `lstmLayer(numHiddenUnits)` creates a Long Short-Term Memory layer. numHiddenUnits is the number of hidden units in the layer. The number of hidden units is the number of neurons in the layer.

3. `fullyConnectedLayer(outputSize)` creates a fully connected layer with specified output size.

4. `regressionLayer` creates a regression output layer for a neural network. Regression is data fitting.

The learning rate starts at 0.005. It is decreased by a factor of 0.2 every 125 epochs in a piecewise manner using these options:

```
'InitialLearnRate',0.005, ...
'LearnRateSchedule','piecewise', ...
'LearnRateDropPeriod',125, ...
'LearnRateDropFactor',0.2, ...
```

We let "patience" be `inf`. This means the learning will continue to the last epoch even if no progress is made. The training window is shown in Figure 10.7. The top plot shows the root-mean-squared error (RMSE) calculated from the data and the bottom plot the loss. We are also using the test data for validation. Note that the validation data needs to be normalized with its own mean and standard deviation.

The final part tests the network using `predictAndUpdateState`. We need to unnormalize the output for plotting.

```
95   %% Demonstrate the neural net
96   yPred    = predict(net,sTest);
97   yPred(1) = yTrain(end-1);
98   yPred(2) = yTrain(end);
99   yPred    = sigma*yPred + mu;
100
101  %% Plot the prediction
102  NewFigure('Stock prediction')
103  plot(t(1:nTrain-1),sTrain(1:end-1));
104  hold on
105  plot(t,s,'--g');
106  grid on
107  hold on
108  k = nTrain+1:n;
```

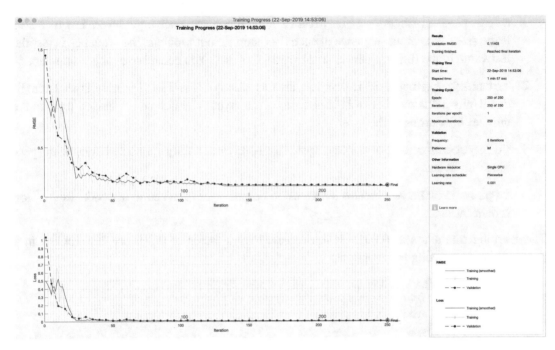

Figure 10.7: *The training window for LSTM with 250 iterations.*

```
109  plot(t(k),[s(nTrain) yPred],'-')
110  xlabel("Year")
111  ylabel("Stock Price")
112  title("Forecast")
113  legend(["Observed" "True" "Forecast"])
114
115  % Format the ticks
116  yT  = get(gca,'YTick');
117  yTL = cell(1,length(yT));
118  for k = 1:length(yT)
119    yTL{k} = sprintf('%5.0f',yT(k));
120  end
121  set(gca,'YTickLabel', yTL)
```

Compare Figure 10.8 with Figure 10.5. The red is the prediction. The prediction reproduces the trend of the stock. It gives you an idea of how it might perform. The neural network cannot predict the exact stock history but does recreate the overall performance that is expected.

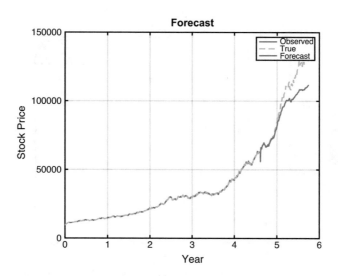

Figure 10.8: *The prediction with one LSTM layer.*

Results for the `bilstm` layer and two `lstm` layers are shown in Figure 10.9 and Figure 10.10. All produce acceptable models. The `bilstm` looks like it predicts the trends better.

This chapter demonstrated that an LSTM can produce an internal model that replicates the behavior of a system from just observations of the process. In this case, we had a model, but in many systems, a model does not exist or has considerable uncertainty in its form. For this reason, neural nets can be a powerful tool when working with dynamical systems. We haven't tried this with real stocks.

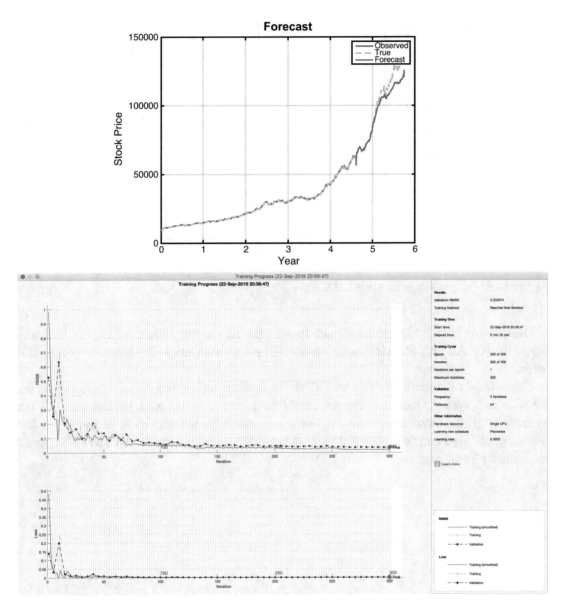

Figure 10.9: *Results for the BiLSTM set.*

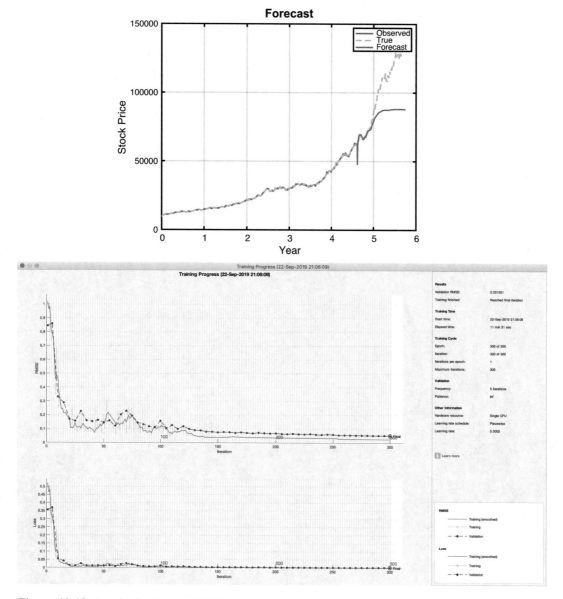

Figure 10.10: *Results for the two LSTM layer set.*

CHAPTER 11

■ ■ ■

Image Classification

11.1　Introduction

Image classification can be done with pretrained networks. MATLAB makes it easy to access and use these networks. This chapter shows you two examples. First, we will use AlexNet, then GoogLeNet.

11.2　Using AlexNet

11.2.1　Problem

We want to use the pretrained network AlexNet for image classification.

11.2.2　Solution

Depending on your version of MATLAB, install AlexNet from the Add-On Explorer or download the support package for GoogLeNet. Load some images and test. These are classification networks, so we will use `classify` to run them.

11.2.3　How It Works

First, we need to download the support packages with the Add-On Explorer. If you attempt to run `alexnet` or `googlenet` without having them installed, you will get a link directly to the package in the Add-On Explorer. You will need your MathWorks password.

AlexNet is a pretrained convolutional neural network (CNN) that has been trained on approximately 1.2 million images from the ImageNet Dataset (`http://image-net.org/index`). The model has 25 layers and can classify images into 1000 object categories. It can be used for all sorts of object classification. However, if an object was not in the training set, it won't be able to identify the object. If a banana was in the training set, you could expect the CNN to correctly identify a new picture of a banana. But if you gave it a picture of a plantain, and plantain was NOT in the CNN, then it might not find a match or, more likely, it might incorrectly classify it like a banana.

© Michael Paluszek, Stephanie Thomas, Eric Ham 2022
M. Paluszek et al., *Practical MATLAB Deep Learning*,
https://doi.org/10.1007/978-1-4842-7912-0_11

AlexNetTest.m

```
 8   %% Load the network
 9   % Access the trained model. This is a SeriesNetwork.
10   net = alexnet;
11   net
12
13   % See details of the architecture
14   net.Layers
```

The network layers printout is shown as follows:

```
>> AlexNetTest

ans =

   25x1 Layer array with layers:

     1   'data'     Image Input                   227x227x3 images with 'zerocenter' normalization
     2   'conv1'    Convolution                   96 11x11x3 convolutions with stride [4  4] and padding [0
                  0  0  0]
     3   'relu1'    ReLU                          ReLU
     4   'norm1'    Cross Channel Normalization   cross channel normalization with 5 channels per element
     5   'pool1'    Max Pooling                   3x3 max pooling with stride [2  2] and padding [0  0  0  0]
     6   'conv2'    Grouped Convolution           2 groups of 128 5x5x48 convolutions with stride [1  1] and
                  padding [2  2  2  2]
     7   'relu2'    ReLU                          ReLU
     8   'norm2'    Cross Channel Normalization   cross channel normalization with 5 channels per element
     9   'pool2'    Max Pooling                   3x3 max pooling with stride [2  2] and padding [0  0  0  0]
    10   'conv3'    Convolution                   384 3x3x256 convolutions with stride [1  1] and padding [1
                  1  1  1]
    11   'relu3'    ReLU                          ReLU
    12   'conv4'    Grouped Convolution           2 groups of 192 3x3x192 convolutions with stride [1  1] and
                  padding [1  1  1  1]
    13   'relu4'    ReLU                          ReLU
    14   'conv5'    Grouped Convolution           2 groups of 128 3x3x192 convolutions with stride [1  1] and
                  padding [1  1  1  1]
    15   'relu5'    ReLU                          ReLU
    16   'pool5'    Max Pooling                   3x3 max pooling with stride [2  2] and padding [0  0  0  0]
    17   'fc6'      Fully Connected               4096 fully connected layer
    18   'relu6'    ReLU                          ReLU
    19   'drop6'    Dropout                       50% dropout
    20   'fc7'      Fully Connected               4096 fully connected layer
    21   'relu7'    ReLU                          ReLU
    22   'drop7'    Dropout                       50% dropout
    23   'fc8'      Fully Connected               1000 fully connected layer
    24   'prob'     Softmax                       softmax
    25   'output'   Classification Output         crossentropyex with 'tench' and 999 other classes
```

There are many layers in this convolutional network. ReLU and softmax are the activation functions. In the first layer, "zero center" normalization is used. This means the images are normalized to have a mean of zero and a standard deviation of one. Two layers are new, Cross Channel Normalization and Grouped Convolution. Filter groups, also known as grouped convolution, were introduced with AlexNet in 2012. You can think of the output of each filter as a channel and filter groups as groups of the channels. Filter groups allowed more efficient parallelization across GPUs. They also improved performance. Cross Channel Normalization normalizes across channels, instead of one channel at a time. We've discussed convolution in Chapter 3. The weights in each filter are determined during training. Dropout is a layer that ignores nodes, randomly, when training the weights. This prevents interdependencies between nodes.

For our first example, we load an image that comes with MATLAB of a set of peppers. This image is larger than the input size of the network, so we use the top-left corner of the image.

Note that each pretrained network has a fixed input image size that we can determine from the first layer.

AlexNetTest.m

```
16  %% Load a test image and classify it
17  % Read the image to classify
18  I = imread('peppers.png');  % ships with MATLAB
19
20  % Adjust size of the image to the net's input layer
21  sz = net.Layers(1).InputSize;
22  I = I(1:sz(1),1:sz(2),1:sz(3));
23
24  % Classify the image using AlexNet
25  [label, scorePeppers] = classify(net, I);
26
27  %  Show the image and the classification results
28  NewFigure('Pepper'); ax = gca;
29  imshow(I);
30  title(ax,label);
31
32  PlotSet(1:length(scorePeppers),scorePeppers,'x label','Category',...
33          'y label','Score','plot title','Peppers');
```

The images and results for the AlexNet example are shown in Figure 11.1. The pepper scores are tightly clustered.

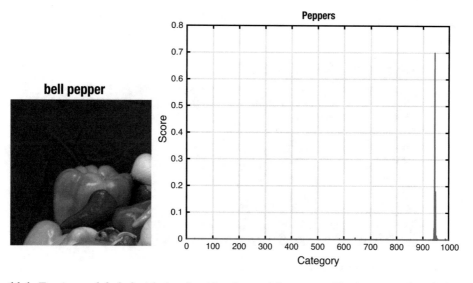

Figure 11.1: *Test image labeled with the classification and the scores. The image is classified as a "bell pepper".*

For fun, and to learn more about this network, we print out the categories that had the next highest scores, sorted from high to low. The categories are stored in the last layer of the net in its Classes.

AlexNetTest.m

```
35  % What other categories are similar?
36  disp('Categories with highest scores for Peppers:')
37  kPos = find(scorePeppers>0.01);
38  [vals,kSort] = sort(scorePeppers(kPos),'descend');
39  for k = 1:length(kSort)
40    fprintf('%13s:\t%g\n',net.Layers(end).Classes(kPos(kSort(k))),vals(k)
        );
41  end
```

The results show that the net was considering all fruits and vegetables. The Granny Smith had the next highest score, followed by cucumber, while the fig and lemon had much smaller scores. This makes sense since Granny Smiths and cucumbers are also usually green.

```
Categories with highest scores for Peppers:

  bell pepper:  0.700013
Granny Smith:  0.180637
    cucumber:  0.0435253
         fig:  0.0144056
       lemon:  0.0100655
```

We also have two of our test images. One is of a cat and one of a metal box, shown in Figure 11.2.

Figure 11.2: *Raw test images Cat.png and Box.jpg.*

The scores for the cat classification are shown as follows:

```
Categories with highest scores for Cat:
            tabby:    0.805644
     Egyptian cat:    0.15372
        tiger cat:    0.0338047
```

The selected label is *tabby*. The net can recognize that the photo is of a cat, as the other highest scored categories are also kinds of cats. Although what a tiger cat might be, as distinguished from a tabby, we can't say...

The metal box proves the biggest challenge to the net. The category scores above 0.05 are shown in the following, and the images with their label are shown in Figure 11.3.

```
Categories with highest scores for Box:
     hard disc:    0.291533
         loupe:    0.0731844
         modem:    0.0702888
          pick:    0.0610284
          iPod:    0.0595867
     CD player:    0.0508571
```

In this case, the hard disc is by far the highest score, but the score is much lower than that of the tabby cat – roughly 0.3 vs. 0.8. The summary of scores is

```
AlexNet results summary:

Pepper         0.7000
Cat            0.8056
Box            0.2915
```

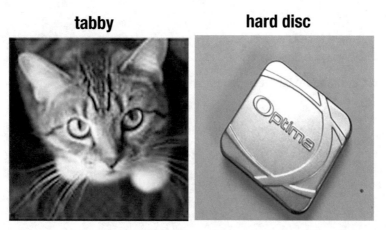

Figure 11.3: *Test images and the classification by AlexNet. They are classified as "tabby" and "hard disc".*

11.3   Using GoogLeNet

11.3.1 Problem

Now let's compare these results to GoogLeNet. GoogLeNet is a pretrained model that has also been trained on a subset of the ImageNet database which is used in the ImageNet Large-Scale Visual Recognition Challenge (ILSVRC). The model is trained on more than a million images, has 144 layers (a lot more than the AlexNet), and can classify images into 1000 object categories.

11.3.2 Solution

Install GoogLeNet from the Add-On Explorer (try running it from the command line to get an install link). Load some images and test using `classify`.

11.3.3 How It Works

First, we load the pretrained network as in the previous recipe.

GoogleNetTest.m

```
10  %% Load the pretrained network
11  net = googlenet;
12  net  % display the 144 layer network
```

The net display is shown as follows. It is a different type than AlexNet, a `DAGNetwork`. This network has its layers arranged as a directed acyclic graph; layers have inputs from and outputs to multiple layers.

```
net =
DAGNetwork with properties:

        Layers: [144x1 nnet.cnn.layer.Layer]
   Connections: [170x2 table]
```

Next, we test it on the image of peppers.

GoogleNetTest.m

```
14  %% The pepper
15  % Read the image to classify
16  I = imread('peppers.png');
17  sz = net.Layers(1).InputSize;
18  I = I(1:sz(1),1:sz(2),1:sz(3));
19  [label, scorePeppers] = classify(net, I);
20  NewFigure('Pepper');
21  imshow(I);
22  title(label);
23  % What other categories are similar?
24  disp('Categories with highest scores for Peppers:');
25  kPos = find(scorePeppers>0.01);
```

```
26   [vals,kSort] = sort(scorePeppers(kPos),'descend');
27   for k = 1:length(kSort)
28     fprintf('%13s:\t%g\n',net.Layers(end).Classes(kPos(kSort(k))),vals(k)
         );
29   end
```

As before, the image is correctly identified as having a bell pepper, and the score is similar to AlexNet. However, the remaining categories are a little different. In this case, the cucumber and, for some reason, a maraca scored higher than a Granny Smith. Maracas are also round and oblong. The highest categories are as follows:

```
Categories with highest scores for Peppers:
  bell pepper:   0.708213
    cucumber:    0.0955994
      maraca:    0.0503938
Granny Smith:    0.0278589
```

We also test this net on the images of the cat and box. The image size for this network is 224×224. The categories for the cat are the same, with the addition of a lynx, and note that the tabby score is significantly lower than for AlexNet.

```
Categories with highest scores for Cat:
          tabby:   0.532261
   Egyptian cat:   0.373229
      tiger cat:   0.0790764
           lynx:   0.0135277
```

The box scores prove the most interesting, and while the hard disc is among the highest scores, in this case, the net returns *iPod*. A cellular telephone is added to the mix this time. The net identifies that it is a rectangular metal object, but beyond that, there is no clear evidence for one category over another.

```
Categories with highest scores for Box:
             iPod:   0.443666
        hard disc:   0.212672
cellular telephone:   0.0787301
            modem:   0.0766429
             pick:   0.0545631
           switch:   0.0169888
            scale:   0.0165957
   remote control:   0.0154203
```

The GoogLeNet score arrays for Cat.png and Box.png are shown in Figure 11.4. The box scores are visibly spread all over the place. This reinforces that the choice of "iPod" is less certain than the pepper or cat. This shows that even highly trained networks are not necessarily reliable if the input strays too far from the test set.

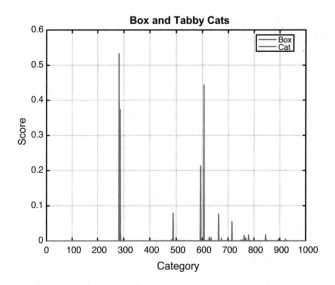

Figure 11.4: *GoogLeNet scores for Cat.png and Box.png.*

```
The summary of the GoogleNet results are:

GoogleNet results summary:

Pepper      0.7082
Cat         0.5323
Box         0.4437
```

We can also grab random images from the Internet. The site picsum.photos calls itself the "Lorem Ipsum" for photos and provides a random photo with every call to the URL. Consider, for example:

```
>> I = imread('https://picsum.photos/224/224');
>> figure, imshow(I);
>> title(classify(net,I))
```

We got some interesting results using this website. It produces good results for some landscape photos, but other times see objects that are not there. Figure 11.5 shows four examples. An overexposed image of a sunset over a train yard is identified as a "volcano." Two landscapes are appropriately labeled as "lakeside" and "seashore." However, in the last image, a person on a bench gazing at a beach or desert landscape is inexplicably identified as "geyser." This may have to do with the shape of the sky or clouds.

These nets are **not** trained on people; however, it can be interesting to test them on images of people. We tested GoogLeNet on our author headshots, Figure 11.6. In both cases, it identified our clothing fairly accurately!

Figure 11.5: *GoogLeNet identification results of random images from* `picsum.photos`.

Figure 11.6: *Author headshots with GoogLeNet labels.*

While these nets perform very well on images that do exist in their database, from lions to landscapes, it is important to remember that they are limited in application. Results can be unexpected and even silly.

CHAPTER 12

■ ■ ■

Orbit Determination

12.1 Introduction

Determining orbits from measurements has been done for hundreds of years. The general approach is to take a series of measurements of the object from the ground at different times. Possible measurements are range and range rate from the observation point or angles of the object from the measurement point. Given the location from where the measurements are taken on the Earth and this set of data, one can reconstruct the orbit. Ideal orbits, which assume that the Earth's gravity is represented by a point mass at the center of the Earth, are conic sections. Those that stay near the Earth are ellipses. These can be defined as a set of orbital elements. In this chapter, we will design a neural network to find the values for two of the elements. Our model will be simpler than that which astronomers must use. We will assume that all of our orbits are in the Earth's equatorial plane and that the observer is at the center of the Earth.

The purpose of this chapter is to show that a neural net can do orbit determination. For comparison with traditional methods, see the classic textbook from 1965 by Escobal [12].

12.2 Generating the Orbits

12.2.1 Problem

We want to create a set of orbits for testing and training a neural net.

12.2.2 Solution

Implement a random orbit generator using Keplerian propagation of elements.

12.2.3 How It Works

An orbit involves at least two bodies, for example, a planet and a spacecraft. In the ideal two-body case, the two bodies rotate about the common center of mass, known as the barycenter. For all practical spacecraft cases, the spacecraft mass itself is negligible, and this means that the

© Michael Paluszek, Stephanie Thomas, Eric Ham 2022
M. Paluszek et al., *Practical MATLAB Deep Learning*,
https://doi.org/10.1007/978-1-4842-7912-0_12

satellite follows a conic section path about the primary body's center of mass. A conic section is a curve that fits on a cone, as shown in Figure 12.1. Two conics, a circle, and an ellipse are drawn. Hyperbolas and parabolas are also conic sections, but we will only look at elliptical orbits in this chapter.

The code that draws this picture is in the following script. It calls two functions, `Cone` and `ConicSectionEllipse`. `r0` and `h` are only needed to draw the cone. The algorithm only cares about `theta`, the half cone angle.

ConicSection.m

```
1  %% Create a conic section
2
3  theta = pi/4;
4  h     = 4;
5  r0    = h*sin(theta);
6
7  ang   = linspace(0,2*pi);
8  a     = 2;
9  b     = 1;
10 cA    = cos(ang);
11 sA    = sin(ang);
12 n     = length(cA);
13 c     = 0.5*h*sin(theta)*[cA;sA;ones(1,n)];
14 e     = [a*cA;b*sA;zeros(1,n)];
15
16 % Show a planar representation
17 NewFigure('Orbits');
18 plot(c(1,:),c(2,:),'r','linewidth',2)
19 hold on
20 plot(e(1,:),e(2,:),'b','linewidth',2)
```

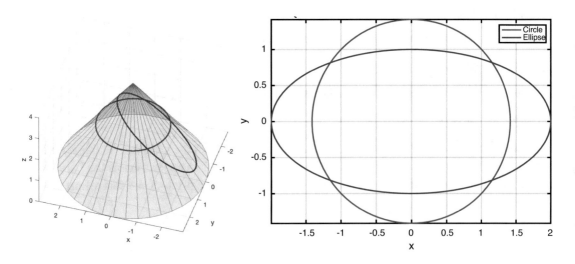

Figure 12.1: *Ellipse and circle on a cone and viewed along their normal.*

```
21  grid on
22  xlabel('x')
23  ylabel('y')
24  axis image
25  legend('Circle','Ellipse');
26
27  [z,phi,x] = ConicSectionEllipse(a,b,theta);
28  ang        = pi/2 + phi;
29  e          = [cos(ang) 0 sin(ang);0 1 0; -sin(ang) 0 cos(ang)]*e;
30  e(1,:)     = e(1,:) + x;
31  e(3,:)     = e(3,:) + h - z;
32
33  Cone(r0,h,40);
34  hold on
35  plot3(c(1,:),c(2,:),2*ones(1,n),'r','linewidth',2);
36  plot3(e(1,:),e(2,:),e(3,:),'b','linewidth',2);
37  view([0 1 0])
```

The view is set to look along the y-axis which is the axis of rotation for the ellipse. The function Cone draws the cone. line draws the axis of rotation that is along the short axis.

The solution that is used to draw the conic sections is derived in the last section of this chapter. The orbit may be elliptical, with an eccentricity less than one, parabolic with an eccentricity equal to one, or hyperbolic with an eccentricity greater than one. Figure 12.2 shows the geometry of an elliptical orbit. This is a planar orbit in which the orbital motion is two-dimensional. The semi-major axis a is

$$a = \frac{r_a + r_p}{2} \tag{12.1}$$

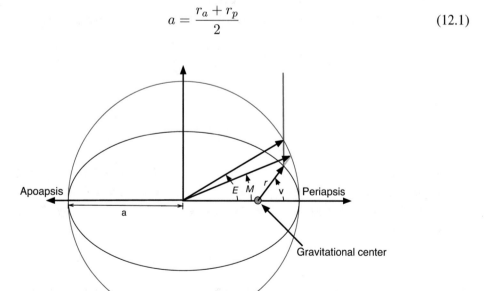

Figure 12.2: *Elliptical orbit.*

237

where r_a is the apoapsis (apogee for the Earth) radius, or point furthest from the central planet, and r_p is the periapsis radius (perigee for the Earth), or closest point to the planet. The eccentricity, e, of the orbit is

$$e = \frac{r_a - r_p}{r_a + r_p} \tag{12.2}$$

When $r_a = r_p$, the orbit is circular and $e = 0$. This formula is not meaningful for parabolic or hyperbolic orbits. Figure 12.2 shows three angular measurements, M mean anomaly, E eccentric anomaly, and ν true anomaly. All are measured from periapsis. The mean anomaly is related to the mean orbit rate n through a simple function of time:

$$M = M_0 + n(t - t_0) \tag{12.3}$$

The eccentric anomaly is the angle to the current position as projected onto the ellipse's circumscribing circle, drawn in blue. It is related to the mean anomaly by Kepler's equation:

$$M = E - e \sin E \tag{12.4}$$

This equation needs to be solved numerically in general, but for small values of e, $e < 0.1$, this approximation can be used:

$$E \approx M + e \sin M + \frac{1}{2} e^2 \sin 2M \tag{12.5}$$

This is because apoapsis is not well defined for very small e. Higher-order formulas can also be found. The true anomaly is related to the eccentric anomaly through the equation

$$\tan \frac{\nu}{2} = \sqrt{\frac{1+e}{1-e}} \tan \frac{E}{2} \tag{12.6}$$

Finally, the orbit radius is

$$r = \frac{a(1 - e)(1 + e)}{1 + e \cos \nu} \tag{12.7}$$

If $e > 1$ in this equation, r will go to ∞, as is expected for parabolic or hyperbolic orbits.

Seven parameters are necessary to define an orbit of a spacecraft about a spherically symmetric body. One is the gravitational parameter, generally denoted by the symbol μ. The gravitational parameter is

$$\mu = G(m_1 + m_2) \tag{12.8}$$

where m_1 is the mass central body and m_2 is the mass of the orbiting body. G is the gravitational constant with units of m³/kg/s². For the Earth, $G = 6.6774 \times 10^{-11}$ m³/kg/s². μ for the Earth is 3.98600436×10^8 m³/s². There are many ways of representing the other six elements. The two most popular sets are position and velocity (r and v) vectors and Keplerian orbital elements.

Each representation uses six independent variables to describe the orbit, plus μ. Both are shown in Figure 12.3.

The Keplerian elements are defined as follows: Two elements define the elliptical orbit. The size of the orbit is determined by the semi-major axis a, which is the average of the perigee radius and apogee radius. The size and shape of the orbit are defined by the eccentricity, e. Two elements define the orbital plane. Ω is the longitude which is the right ascension of the ascending node, or the angle from the $+x_{ECI}$ axis of the reference frame to the line where the orbit plane intersects the xy-plane. i is the inclination and is the angle between the $x_{ECI}y_{ECI}$-plane and the orbit plane. ω is the argument of perigee and is the angle in the orbit plane between the ascending node line and perigee (where the orbit is closest to the center of the central body). ν is the true anomaly and is the angle between the perigee and the spacecraft. The mean anomaly M may be used in the element set instead of ν. M or ν tells us where the spacecraft is in its orbit. To summarize, the Keplerian elements are

$$x = \begin{bmatrix} a \\ i \\ \Omega \\ \omega \\ e \\ M \end{bmatrix} \tag{12.9}$$

The orbit period, with units of seconds, is

$$P = 2\pi\sqrt{\frac{a^3}{\mu}} \tag{12.10}$$

The orbit parameter, with units of distance (conventionally km), is

$$p = a(1 - e)(1 + e) \tag{12.11}$$

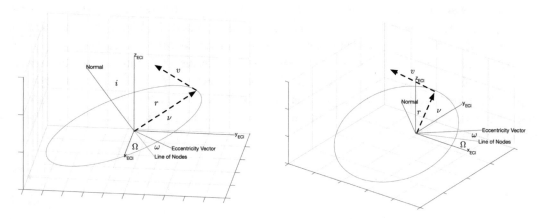

Figure 12.3: *Orbital elements from two vantage points. The underlying plot was drawn using* DrawEllipticOrbit.

The in-plane position and velocity are

$$r_p = \frac{p}{1 + e \cos \nu} \begin{bmatrix} \cos \nu \\ \sin \nu \\ 0 \end{bmatrix} \tag{12.12}$$

$$v_p = \sqrt{\frac{\mu}{p}} \begin{bmatrix} -\sin \nu \\ e + \cos \nu \\ 0 \end{bmatrix} \tag{12.13}$$

The transformation matrix from planar to three-dimensional coordinates is

$$c = \begin{bmatrix} \cos \Omega \cos \omega - \sin \Omega \sin \omega \cos i & -\cos \Omega \sin \omega - \sin \Omega \cos \omega \cos i & \sin \Omega \sin i \\ \sin \Omega \cos \omega + \cos \Omega \sin \omega \cos i & -\sin \Omega \sin \omega + \cos \Omega \cos \omega \cos i & -\cos \Omega \sin i \\ \sin \omega \sin i & \cos \omega \sin i & \cos i \end{bmatrix} \tag{12.14}$$

That is

$$r = cr_p \tag{12.15}$$

$$v = cv_p \tag{12.16}$$

For the purposes of creating the neural net, we will look at orbits with the inclination, $i = 0$, and the ascending node, $\Omega = 0$. The transformation matrix reduces to a rotation about z.

$$c = \begin{bmatrix} \cos \omega & -\sin \omega & 0 \\ \sin \omega & \cos \omega & 0 \\ 0 & 0 & 1 \end{bmatrix} \tag{12.17}$$

We now want to propagate the orbit forward in time. There are two alternative approaches to doing so. One approach is to use Keplerian propagation, where we keep five of the elements constant, and simply march the mean anomaly forward in time at a constant rate of $n = \sqrt{\mu/a^3}$. At each point in time, we can convert the set of six orbital elements into a new position and velocity. This approach is limited though, in that it assumes the orbit follows a Keplerian orbit. The second approach, which gives us more flexibility, for external forces like thrust and drag, is to numerically integrate the dynamic equations of motion. The state equations for orbit propagation are

$$\dot{v} = -\mu \frac{r}{|r|^3} + a \tag{12.18}$$

$$\dot{r} = v \tag{12.19}$$

The terms on the right-hand side of the velocity derivative equation are the point-mass gravity acceleration with additional acceleration a. This is implemented in RHSOrbit.

RHSOrbit.m

```
16   function xDot = RHSOrbit(~,x,d)
17
18   r      = x(1:2);
19   v      = x(3:4);
20   xDot   = [v;-d.mu*r/(r'*r)^1.5 + d.a];
```

We will create a script that simulates multiple orbits. The simulation will use RHSOrbit. The first part of the orbit generation script sets up the random orbital elements.

Orbits.m

```
1    %% Generate Orbits for angles-only element estimation
2    % Saves a mat-file called OrbitData.
5
6    nEl    = 500;              % Number of sets of data
7    d      = struct;           % Initialize
8    d.mu   = 3.98600436e5;     % Gravitational parameter, km^3/s^2
9    d.a    = [0;0];            % Perturbing acceleration
10
11   % Random elements
12   e      = 0.6*rand(1,nEl);              % Eccentricity
13   a      = 8000 + 1000*randn(1,nEl);     % Semi-major axis
14   M      = 0.25*pi*rand(1,nEl);          % Mean anomaly
```

The next section runs the simulations and saves the angles. Each simulation has 2000 steps, and each step is two seconds. We are only using one in ten points for the orbit determination. We save the orbital elements for testing the neural network. We are not applying any external acceleration. We could have used Kepler propagation, but by simulating the orbit, we have the option of studying how well the neural network performs with disturbances.

```
16   % Set up the simulation
17   nSim   = 2000; % Number of simulation steps
18   dT     = 2; % Time step
19
20   % Only use some of the sim steps
21   jUse   = 1:10:nSim;
22
23   % Data for Deep Learning
24   data   = cell(nEl,1);
25
26   %% Simulate each of the orbits
27   x      = zeros(4,nSim);
28   t      = (0:(nSim-1))*dT;
29   el(nEl) = struct('a',7000,'e',0); % initialize struct array
30
31   for k = 1:nEl
32     [r,v] = El2RV([a(k) 0 0 0 e(k) M(k)]);
33     x     = [r(1:2);v(1:2)];
34     xP    = zeros(4,nSim);
```

```
35    for j = 1:nSim
36      xP(:,j) = x;
37      x       = RungeKutta( @RHSOrbit, 0, x, dT, d );
38    end
39    data{k} = atan2(xP(2,jUse),xP(1,jUse));
40    el(k).a = a(k);
41    el(k).e = e(k);
42  end
```

The final part plots the orbits and saves the data to a file.

```
44  %% Save for the Deep Learning algorithm
45  save('OrbitData','data','el');
46
47  %% Plot
48  [t, tL] = TimeLabel(t(jUse));
49  angle   = data{k}(1,:);
50  PlotSet(t,angle,'x label', tL,'y label','Angle (rad)','figure title','
        Angle');
51  PlotSet(xP(1,:),xP(2,:),'x label', 'x (km)','y label','y (km)','figure
        title','Orbit');
```

The last orbit is shown in Figure 12.4. The jump in angle is due to angles being defined from $-\pi$ to $+\pi$. We could have used unwrap to get rid of this jump. We are only measuring for the part of an orbit. We can set up the simulation to measure any part of an orbit or even multiple orbits.

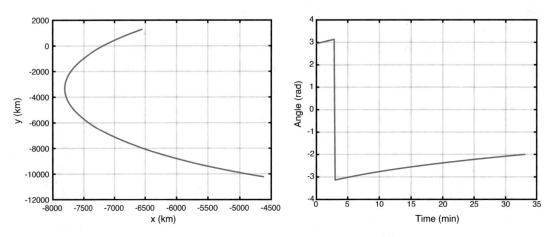

Figure 12.4: *The last test orbit. The measured angle is on the right. These are only showing the data used in orbit determination.*

12.3   Training and Testing

12.3.1 Problem

We want to build a deep learning system to compute the eccentricity and semi-major axis for an orbit from angle measurements. The argument of perigee will be assumed to be zero.

12.3.2 Solution

The orbit history is a time series of angles. We will take angles at uniform time intervals. We will use `fitnet` to fit the data, which creates a function-fitting neural network with a specified hidden layer size and training function.

12.3.3 How It Works

We load in the data from the mat-file and separate it into training and testing sets.

OrbitNeuralNet.m

```matlab
1    %% Train and test the Orbit Neural Net

4
5    s        = load('OrbitData');
6    n        = length(s.data);
7    nTrain   = floor(0.9*n);

8
9    %% Set up the training and test sets
10   kTrain   = randperm(n,nTrain);
11   sTrain   = s.data(kTrain);
12   nSamp    = size(sTrain{1},2);
13   xTrain   = zeros(nSamp,nTrain);
14   aMean    = mean([s.el(:).a]);

15
16   for k = 1:nTrain
17     xTrain(:,k) = sTrain{k}(1,:);
18   end

19
20   elTrain      = s.el(kTrain);
21   yTrain       = [elTrain.a;elTrain.e];
22   yTrain(1,:)  = yTrain(1,:)/aMean; % Normalize the data
23   kTest        = setdiff(1:n,kTrain);
24   sTest        = s.data(kTest);
25   nTest        = n-nTrain;
26   xTest        = zeros(nSamp,nTest);
27   for k = 1:nTest
28     xTest(:,k) = sTest{k}(1,:);
29   end

30
31   elTest       = s.el(kTest);
32   yTest        = [elTest.a;elTest.e];
33   yTest(1,:)   = yTest(1,:)/aMean;
```

243

The neural network will use sequences of angles and their related times as the input. The output will be the two orbital elements: semi-major axis and eccentricity. In general, if we know the position and velocity at a point in the orbit, we can always compute the orbital elements. This is done in the function El2RV. The neural network, in effect, will infer position and velocity just from the angles.

We train the network using fitnet and later in the script compare that to cascadeforw ardnet and feedforwardnet. Note that we normalized the semi-major axis so that the magnitude is in the same order as the eccentricity. This improves the fitting.

```
35  %% Train the network
36  net         = fitnet(10);
37
38  net         = configure(net, xTrain, yTrain);
39  net.name    = 'Orbit';
40  net         = train(net,xTrain,yTrain);
```

We use the test data to test the network.

```
42  %% Test the network
43  yPred       = sim(net,xTest);
44  yPred(1,:)  = yPred(1,:)*aMean;
45  yTest(1,:)  = yTest(1,:)*aMean;
46  yM          = mean(yPred-yTest,2);
47  yTM         = mean(yTest,2);
48  fprintf('\nFit Net\n');
49  fprintf('Mean semi-major axis error %12.4f (km) %12.2f %%\n',yM(1),100*
        abs(yM(1))/yTM(1));
50  fprintf('Mean eccentricity    error %12.4f      %12.2f %%\n',yM(2),100*
        abs(yM(2))/yTM(2));
51
52  %% Plot the results
53  NewFigure('Predictions using Fitnet')
54  subplot(2,1,1)
55  bar(1:nTest,[yPred(1,:);yTest(1,:)]);
56  ylabel('a')
57  legend('Predicted','True')
58  subplot(2,1,2)
59  bar(1:nTest,[yPred(2,:);yTest(2,:)]);
60  ylabel('e')
61  legend('Predicted','True')
```

The results are best for fitnet. However, the results will vary with each run.

```
>> OrbitNeuralNet

Fit Net
Mean semi-major axis error      31.9872 (km)       0.41 %
Mean eccentricity    error       0.0067             2.48 %
```

```
Cascade Forward Net
Mean semi-major axis error     -89.8603 (km)         1.15 %
Mean eccentricity    error      -0.0100              3.74 %


Feed Forward Net
Mean semi-major axis error      40.2986 (km)         0.52 %
Mean eccentricity    error       0.0001              0.03 %
```

Figure 12.5, Figure 12.6, and Figure 12.7 show the test results. Both semi-major axis and eccentricity results are reasonably good. You can experiment with different spans of data and different sampling intervals.

We then train the network using `cascadeforwardnet`. The code doesn't change except for the function name.

```
64  %% Train the cascade forward network
65  net        = cascadeforwardnet(10);
66  net        = configure(net, xTrain, yTrain);
67  net.name   = 'Orbit';
68  net        = train(net,xTrain,yTrain);
```

We finally train it using `feedforwardnet`.

```
92  %% Train the feed forward network
93  net       = feedforwardnet(10);
94  net       = configure(net, xTrain, yTrain);
95  net.name  = 'Orbit';
96  net       = train(net,xTrain,yTrain);
```

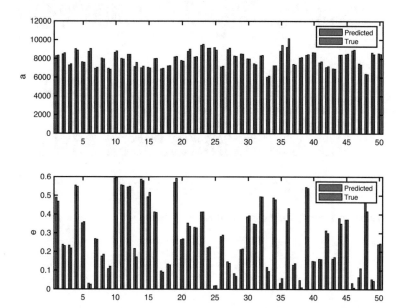

Figure 12.5: *Test results using fitnet.*

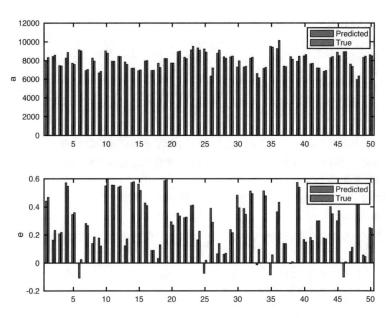

Figure 12.6: *Test results using* `cascadeforwardnet`.

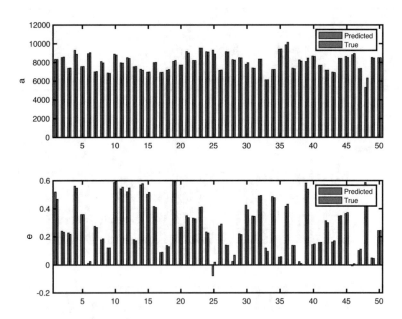

Figure 12.7: *Test results using* `feedforwardnet`.

Both the feedforward network and cascade-forward network produce negative eccentricities when the true eccentricities are near zero.

12.4   Implementing an LSTM

12.4.1 Problem

We want to build a Long Short-Term Memory neural net (LSTM) to estimate the orbital elements. LSTMs have been demonstrated in previous chapters. An LSTM can learn from sequences of measurements, which should be advantageous in orbit determination. They are an alternative to the functions shown earlier.

12.4.2 Solution

The orbit history is a time series of angles. We will use a bidirectional LSTM to fit the data. We will take angles at uniform time intervals.

12.4.3 How It Works

The script is `OrbitLSTM.m`. We load in the data from the mat-file and separate it into training and testing sets. The data format is different from the feedforward networks. `xTrain` is a cell array, but `yTrain` is a matrix with a row for each cell array in `xTrain`.

OrbitLSTM.m

```matlab
1  %% Script to train and test the Orbit LSTM
2  % It will estimate the orbit semi-major axis and eccentricity from a
3  % time sequence of angle measurements.
8  s              = load('OrbitData');
9  n              = length(s.data);
10 nTrain         = floor(0.9*n);
11
12 %% Set up the training and test sets
13 kTrain         = randperm(n,nTrain);
14 aMean          = mean([s.el(:).a]);
15 xTrain         = s.data(kTrain);
16 nTest          = n-nTrain;
17
18 elTrain        = s.el(kTrain);
19 yTrain         = [elTrain.a;elTrain.e]';
20 yTrain(:,1)    = yTrain(:,1)/aMean;
21 kTest          = setdiff(1:n,kTrain);
22 xTest          = s.data(kTest);
23
24 elTest         = s.el(kTest);
25 yTest          = [elTest.a;elTest.e]';
26 yTest(:,1)     = yTest(:,1)/aMean;
```

We train the network using `trainNetwork`.

```
28  %% Train the network with validation
29  numFeatures     = 1;
30  numHiddenUnits1 = 100;
31  numHiddenUnits2 = 100;
32  numClasses      = 2;
33
34  layers = [ ...
35      sequenceInputLayer(numFeatures)
36      bilstmLayer(numHiddenUnits1,'OutputMode','sequence')
37      dropoutLayer(0.2)
38      bilstmLayer(numHiddenUnits2,'OutputMode','last')
39      fullyConnectedLayer(numClasses)
40      regressionLayer]
41
42  maxEpochs = 20;
43
44  options = trainingOptions('adam', ...
45      'ExecutionEnvironment','cpu', ...
46      'GradientThreshold',1, ...
47      'MaxEpochs',maxEpochs, ...
48      'Shuffle','every-epoch', ...
49      'ValidationData',{xTest,yTest}, ...
50      'ValidationFrequency',5, ...
51      'Verbose',0, ...
52      'Plots','training-progress');
53
54  net = trainNetwork(xTrain,yTrain,layers,options);
```

`options` is given validation data (the same as the test data in this, for simplicity). Note the cell array that is required for the validation data.

```
      'ValidationData',{xTest,yTest}, ...
      'ValidationFrequency',5, ...
```

We shuffle the data. This generally improves the results since the learning algorithm sees the data in a different order on each epoch. We use the test data to test the network. `predict` shows the results produced by the test data. This is the same data used for validation during learning.

```
56  %% Test the network
57  yPred       = predict(net,xTest);
58  yPred(:,1)  = yPred(:,1)*aMean;
59  yTest(:,1)  = yTest(:,1)*aMean;
60  yM          = mean(yPred-yTest,1);
61  fprintf('\nbiLSTM\n');
62  fprintf('Mean semi-major axis error %12.4f (km)\n',yM(1));
63  fprintf('Mean eccentricity    error %12.4f\n',yM(2));
```

The results are given as follows:

```
>> OrbitLSTM
layers =
  6x1 Layer array with layers:

     1   ''    Sequence Input       Sequence input with 1 dimensions
     2   ''    BiLSTM               BiLSTM with 100 hidden units
     3   ''    Dropout              20% dropout
     4   ''    BiLSTM               BiLSTM with 100 hidden units
     5   ''    Fully Connected      2 fully connected layer
     6   ''    Regression Output    mean-squared-error

biLSTM
Mean semi-major axis error       -63.4780 (km)
Mean eccentricity    error         0.0024
```

We use two BiLSTM layers with a 20% dropout between layers. Dropout removes neurons and helps prevent overfitting. Overfitting is when the results correspond too closely to a particular set of data. This makes it hard for the trained network to identify patterns in new data. The first BiLSTM layer produces a sequence as its output. The second BiLSTM layer's `'OutputMode'` is set to `'last'`. The `numClasses` is 2 because we are estimating two parameters. The fully connected layer connects the two BiLSTM outputs to the two parameters we want to identify in the regression layer. The training window is shown in Figure 12.8. We

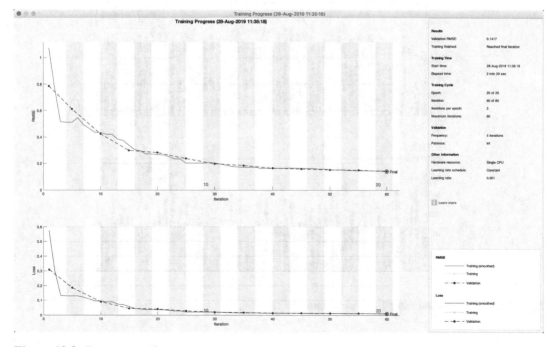

Figure 12.8: *Training window.*

249

could have continued the training for more epochs as the root-mean-squared error (RMSE) is still improving.

This particular set of layers is to show you how to build a neural network. It is by no means the "best" architecture for this problem. We did try a single LSTM layer but found that a single BiLSTM layer worked better.

Figure 12.9 shows the test results. The results are not quite as good as the feedforward nets given earlier. We've only used two layers. From Chapter 11, you see that networks deployed in real applications can have dozens if not hundreds of layers. The difference is due to the smaller number of neurons in the LSTM. You can experiment with this network to improve the results.

In this chapter, we have compared two approaches to solving the orbit determination problem. Using the MATLAB feedforward network worked a bit better than the LSTM network that we implemented. We fixed the argument of perigee to make the problem easier. The next step would be to try and find the full set of orbital elements and then try to design a system that works from a fixed point on the Earth. In the latter case, we would need to account for the rotation of the Earth. Another improvement would be to take the measurements at different time steps. For an elliptical orbit, taking many measurements at perigee is more productive than at apogee because the spacecraft is moving faster. One could write a preprocessor to select inputs to our neural network based on the angular change with respect to time. Orbit determination systems, using algorithmic approaches, can also compute errors in the observer's location. You could also try other measurements, such as range and range rate. These measurements are used for deep space and geosynchronous spacecraft.

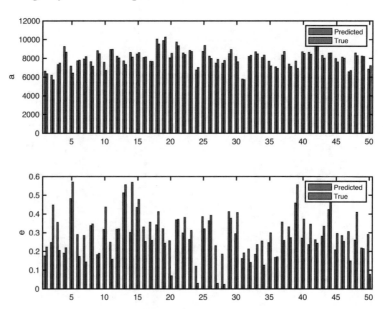

Figure 12.9: *Test results using the bidirectional LSTM.*

CHAPTER 13

■ ■ ■

Earth Sensors

13.1   Introduction

Earth sensors were some of the earliest spacecraft sensors. Many early satellites were for communications or Earth observations, so the most important thing was to point the satellite at the Earth, that is, to null the roll and pitch errors. In many cases, the yaw error was not important or could be controlled through the momentum of the satellite. Earth sensors were either static, without moving parts, or scanning. Scanning was done by rotating the satellite, rotating a mirror with a motor, or rotating a mirror with a torsion bar. Most sensors operated in the infrared wavelengths because the radiation at those wavelengths was more uniform than that in the visible bands and is unaffected by eclipses. Many CubeSats use static Earth sensors because they are simple, compact, and inexpensive. A wide variety of models are available.

This chapter will look at how neural networks can be used to generate roll and pitch angles from any static Earth sensor. Figure 13.1 shows the attitude (also known as orientation) geometry for small angles from ideal Earth pointing. Roll is about the x-axis and pitch is about the y-axis.

Figure 13.2 shows how a static Earth sensor operates. The image on the left shows the Earth sensor elements, represented by the circles centered on the edge of the Earth. The roll and pitch angles are zero. Each element sees part of the Earth and part of space so they all produce the same signal. The picture on the right shows the sensor with a roll and pitch angle. Some elements see just space, some see just the Earth, and some both the Earth and space, but their output will be different from the zero angle cases. By comparing the outputs of the sensors, we get a measurement of roll and pitch. The sensors see light in the CO_2 band, around 14 microns wavelength, so the Earth looks uniform. The elements produce a voltage proportional to their temperature.

Static Earth sensors from the 1960s had built-in logic to go from sensor outputs to roll and pitch. Given the hardware limitations, the algorithms were quite simple. We will develop a neural net that can do this job.

© Michael Paluszek, Stephanie Thomas, Eric Ham 2022   251
M. Paluszek et al., *Practical MATLAB Deep Learning*,
https://doi.org/10.1007/978-1-4842-7912-0_13

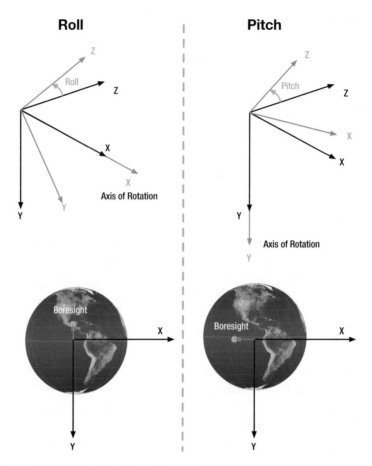

Figure 13.1: *Attitude geometry showing roll and pitch.*

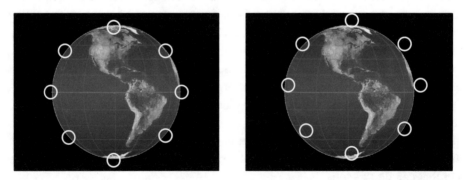

Figure 13.2: *Static Earth sensor operation. The circles represent the sensor elements. The left shows the geometry with zero roll and pitch. The right shows the geometry with nonzero roll and pitch angles. When the angle changes, the portion of the sensor that sees the Earth changes.*

13.2   Linear Output Earth Sensor

13.2.1 Problem

We want to model the output of a linear output, multi-element Earth sensor. The sensor has multiple thermal sensors that produce an output based on their temperature. The maximum value is when the sensor only sees the Earth. The minimum value is when the sensor sees space.

13.2.2 Solution

Use MATLAB's `polyshape` and `intersect` functions to model the elements.

13.2.3 How It Works

The function we create is `StaticEarthSensor`. If there are no inputs, the first part of the function either runs a built-in demo or returns the default data structure. It runs the demo if no outputs are requested.

StaticEarthSensor.m

```
 1   %% STATICEARTHSENSOR Static Earth sensor model
30   function y = StaticEarthSensor(q,r,d)
31
32   % Demo
33   if( nargin < 1 )
34     if( nargout == 0 )
35       Demo
36     else
37       y = DefaultDataStructure;
38     end
39     return
40   end
```

The code transforms the Earth's location into the focal plane of the sensor. It assumes a pinhole camera. The code takes an input attitude quaternion and converts it into locations of the sensor elements. A quaternion is a four-element set that defines the orientation of one frame of reference with respect to another frame of reference. Quaternions are just one way of representing orientation. Three-by-three (3×3) rotation matrices and Euler angles are other ways to represent attitude. Chapter 7 gives additional information on quaternions.

StaticEarthSensor.m

```
42   % Quaternion from ECI to the sensor frame
43   qECIToSensor  = QMult(q,d.qBodyToSensor);
44
45   uNadir        = QForm(qECIToSensor,Unit(r));
46   cEarth        = d.fL*uNadir(1:2);
47
48   % Planet angular width
```

```
49  ang              = asin(d.rPlanet/Mag(r));
50  r                = d.fL*sin(ang);
51
52  % Cycle through all the elements
53  nP               = length(d.az);
54  poly             = cell(1,nP);
55  a                = zeros(1,nP);
56
57  if( d.pAng(1) == 0 )
58    p  = d.p;
59  else
60    p  = d.fL*SinD(d.pAng);
61  end
62
63  for k = 1:nP
64    az               = d.az(k);
65    el               = d.el(k);
66    rP               = d.fL*[cos(az);sin(az)]*sin(el);
67    c                = [cos(az) -sin(az);sin(az) cos(az)];
68    pK               = rP + c*p;
69    [a(k), poly{k}]  = EarthSensorElement(r,d.n,pK,cEarth);
70  end
71
72  % Convert from area to output (linear model)
73  y = a/d.scale;
```

The last line scales the area, producing the output that is linear to the illuminated area.

If no outputs are requested, it draws the sensor. The center circle is the planet, and the sensor elements are superimposed. The sensor elements can be any polygon.

StaticEarthSensor.m

```
75  % Default output
76  if( nargout < 1 )
77    r      = Mag(r);
78    a      = linspace(0,2*pi-2*pi/d.n,d.n);
79    x      = r*cos(a)  + cEarth(1);
80    y      = r*sin(a)  + cEarth(2);
81    planet = polyshape(x,y);
82
83    NewFig('Earth Sensor')
84    plot(planet)
85    hold on
86    for k = 1:nP
87      plot(poly{k})
88    end
89    grid on
90    axis image
91  end
```

The function has a default data structure to help the user. It is for an eight-element sensor.

StaticEarthSensor.m

```
99   %% StaticEarthSensor>DefaultDataStructure
100  function d = DefaultDataStructure
101
102  d.n             = 40;
103  d.p             = [1 1 -1 -1;1 -1 -1 1];
104  d.pAng          = zeros(2,4);
105  d.az            = 0:pi/4:2*pi-pi/4;
106  d.el            = (pi/20)*ones(1,8);
107  d.rPlanet       = 6378.165;
108  d.fL            = 50;
109  d.qBodyToSensor = [1;0;0;0];
110  d.scale         = 1;
```

The code that produces the outputs is in the subfunction `EarthSensorElement`. This uses `polyshape` and `intersect`. The polygon class is a very useful way to represent polygons. MATLAB has many functions that work on polygons.

StaticEarthSensor.m

```
112  %% StaticEarthSensor>EarthSensorElement
113  function [a,poly1] = EarthSensorElement(r,n,p,c)
114
115  poly1 = polyshape(p(1,:),p(2,:));
116  a     = linspace(0,2*pi-2*pi/n,n);
117
118  x     = r*cos(a) + c(1);
119  y     = r*sin(a) + c(2);
120
121  poly2 = polyshape(x,y);
122  poly3 = intersect(poly1,poly2);
123
124  a     = area(poly3);
```

The built-in demo produces an eight-element static Earth sensor, as seen by the number of elements in the `az` and `el` fields in the default data.

StaticEarthSensor.m

```
93   %% StaticEarthSensor>Demo
94   function Demo
95   d   = StaticEarthSensor;
96   r   = [0;0;42167];
97   StaticEarthSensor([1;0;0;0],r,d);
```

The Earth sensor is shown in Figure 13.3.

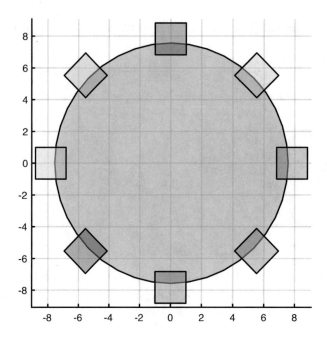

Figure 13.3: *Static Earth sensor from the built-in demo.*

13.3 Segmented Earth Sensor

13.3.1 Problem

We want to model the output of a segmented multi-element Earth sensor. As seen in Figure 13.4, this means that each edge-measuring sensing element is an array instead of a single sensor as in the previous recipe; this allows the edge of the Earth to be localized along with the array.

13.3.2 Solution

Use MATLAB's `polyshape` and `intersect` functions to model the elements.

13.3.3 How It Works

The function used for the multi-element sensor, `SegmentedEarthSensor`, is very similar to the linear sensor model. We'll still describe all of the code. The first part of the function either runs a built-in demo or returns the default data structure.

SegmentedEarthSensor.m

```
1   %% SEGMENTEDEARTHSENSOR Earth sensor model
28  function y = SegmentedEarthSensor(q,r,d)
29
30  % Demo
31  if( nargin < 1 )
32    if( nargout == 0 )
33      Demo
```

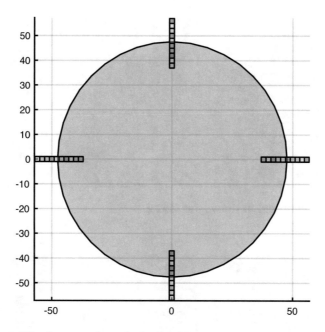

Figure 13.4: *Segmented Earth sensor from the built-in demo.*

```
34    else
35       y = DefaultDataStructure;
36    end
37    return
38  end
```

The body of the function transforms the Earth's location into the focal plane of the sensor. It assumes a pinhole camera. The code takes an input attitude quaternion and converts it into locations of the sensor elements. If a segment is illuminated at all, it returns a one for that element. Otherwise, it returns zero.

SegmentedEarthSensor.m

```
40  qECIToSensor   = QMult(q,d.qBodyToSensor);
41
42  uNadir         = QForm(qECIToSensor,Unit(r));
43  cEarth         = d.fL*uNadir(1:2);
44
45  % Planet angular width
46  ang            = asin(d.rPlanet/Mag(r));
47  r              = d.fL*sin(ang);
48
49  % Cycle through all the elements
50  nP             = length(d.az);
51  poly           = cell(1,nP);
52  y              = zeros(1,nP);
```

```
53
54  for k = 1:nP
55     az               = d.az(k);
56     el               = d.el(k);
57     rP               = [cos(az);sin(az)]*el;
58     c                = [cos(az) -sin(az);sin(az) cos(az)];
59     pK               = rP + c*d.p;
60     [y(k), poly{k}]  = EarthSensorElement(r,d.n,pK,cEarth);
61  end
62
63  y(y > 0) = 1;
```

If no outputs are requested, it draws the sensor.

SegmentedEarthSensor.m

```
65  % Default output
66  if( nargout < 1 )
67     r        = Mag(r);
68          a           = linspace(0,2*pi-2*pi/d.n,d.n);
69     x        = r*cos(a) + cEarth(1);
70     y        = r*sin(a) + cEarth(2);
71     planet   = polyshape(x,y);
72
73     NewFig('Earth Sensor')
74     plot(planet)
75     hold on
76     for k = 1:nP
77       plot(poly{k})
78     end
79     grid on
80     axis image
81  end
```

The function has a default data structure to help the user. It is for four ten-element sensors.

SegmentedEarthSensor.m

```
90  function d = DefaultDataStructure
91
92  d.n             = 40;
93  d.nSeg          = 10;
94  d.p             = [  1      1      -1     -1
95                       1     -1      -1      1];
96  d.az            = [zeros(1,10) (pi/2)*ones(1,10) pi*ones(1,10), (3*pi
        /2)*ones(1,10)];
97  d.fL            = 50;
98  el              = (0:2:18) + d.fL - 12;
99  d.el            = [el el el el];
100 d.rPlanet       = 6378.165;
101 d.qBodyToSensor = [1;0;0;0];
```

Again, the core code that produces the outputs is in the subfunction `EarthSensorEle`ment. This uses `polyshape` and `intersect`.

SegmentedEarthSensor.m

```
103  %% StaticEarthSensor>EarthSensorElement
104  function [a,poly1] = EarthSensorElement(r,n,p,c)
105
106  poly1 = polyshape(p(1,:),p(2,:));
107
108  a    = linspace(0,2*pi-2*pi/n,n);
109  x    = r*cos(a) + c(1);
110  y    = r*sin(a) + c(2);
111
112  poly2 = polyshape(x,y);
113  poly3 = intersect(poly1,poly2);
114
115  a    = area(poly3);
```

The built-in demo produces a four-set, ten-element segmented Earth sensor.

SegmentedEarthSensor.m

```
83  %% StaticEarthSensor>Demo
84  function Demo
85  d   = SegmentedEarthSensor;
86  r   = [0;0;6700];
87  SegmentedEarthSensor([1;0;0;0],r,d);
```

The segmented Earth sensor is shown in Figure 13.4.

13.4   Linear Output Sensor Neural Network

13.4.1 Problem

We want a model that can convert linear sensor outputs from multiple detectors to roll and pitch measurements. The sensors are the radiation detectors that see the edge of the Earth.

13.4.2 Solution

Use a feedforward neural network.

13.4.3 How It Works

The first part of the script, `NNEarthSensor.m`, generates the training data. It calls `ISS Orbit` to get the Keplerian orbital elements for the International Space Station (ISS). It then gets the default data structure from `StaticEarthSensor`. The script then converts orbital elements into position and velocity vectors using `RVOrbGen` and computes the quaternion from the Earth-Centered Inertial Frame (ECI) to the Local Vertical Local Horizontal (LVLH) which is the normal Earth-pointing attitude.

NNEarthSensor.m

```
1   %% Demonstrate LEO static Earth Sensor using a neural network.
2   % The neural network is trained using known roll and pitch.
3
4   degToRad      = pi/180;
5   rE            = Constant('equatorial radius earth');
6   [el,jD0]      = ISSOrbit;
7   d             = StaticEarthSensor;
8   [r,v,t]       = RVOrbGen(el);
9   rMean         = mean(Mag(r));
10  qECIToLVLH    = QLVLH(r,v);
11  d.el          = 64*ones(1,4)*degToRad;
12  d.az          = [0 pi/2 pi 3*pi/2] + pi/4;
13  d.pAng        = 4*[  1      1      -1      -1
14                       1     -1      -1       1];
15
16  n             = 20;
17  roll          = linspace(-6,6,n);
18  pitch         = linspace(-6,6,n);
19  i             = 1;
20  y             = zeros(4,n*n);
21  x             = zeros(2,n*n);
22
23  StaticEarthSensor(qECIToLVLH(:,1),r(:,1),d)
24
25  for j = 1:n
26    for k = 1:n
27      rJ              = roll(j);
28      pK              = pitch(k);
29      mRoll           = [1 0 0;0 CosD(rJ) -SinD(rJ);0 SinD(rJ) CosD(rJ)];
30      mPitch          = [CosD(pK) 0 -SinD(pK);0 1 0;SinD(pK) 0 CosD(pK)];
31      qLVLHToBody     = Mat2Q(mRoll*mPitch);
32      qECIToBody      = QMult(qECIToLVLH(:,1),qLVLHToBody);
33
34      y(:,i)          = StaticEarthSensor(qECIToBody,r(:,1),d);
35      x(:,i)          = [roll(j);pitch(k)];
36      i               = i + 1;
37    end
38  end
```

We then create a four-element Earth sensor model, replacing the default fields in d. This line of code draws the sensor:

NNEarthSensor.m

```
23  StaticEarthSensor(qECIToLVLH(:,1),r(:,1),d)
```

The remaining code creates the training data by inputting different roll and pitch angles and saving the resulting sensor element values in y. The second part trains the feedforward neural network. You first create the neural network data structure. You then configure it by passing it the inputs and outputs. train trains the network. sim simulates the neural network. The variable net is updated with each function call.

NNEarthSensor.m

```
40  % Neural net training
41  net = feedforwardnet(20); % Geneate the neural net structure
42  net = configure( net, y, x ); % Configure based on inputs and outputs
43
44  net.layers{1}.transferFcn = 'poslin'; % Set as purelin
45  net.name  = 'Earth Sensor';
46  net       = train(net,y,x); % Train the network
47  c         = sim(net,y); % Simulate the neural network
```

The next few lines plot the inputs and simulation outputs.

NNEarthSensor.m

```
48  leg       = {'True' 'Neural Net'};
49
50  PlotSet(1:size(c,2),[x;c],'x label','Set',...
51    'y label',{'Roll' 'Pitch'},'figure title','Neural Network',...
52    'plot set',{[1 3],[2 4]}'legend',{leg leg});
53
54  yL = {'Roll' 'Pitch' 'y_1' 'y_2' 'y_3' 'y_4'};
55  PlotSet(1:size(c,2),[x;y],'x label','Set','y label',yL,'Neural Network
        Data')
```

The last part tests the neural network.

NNEarthSensor.m

```
57  %% Testing
58  n       = length(t);
59  roll    = 2;
60  pitch   = 0;
61  c       = zeros(2,n);
62  for k = 1:n
63    rJ            = roll;
64    pK            = pitch;
65    mRoll         = [1 0 0;0 CosD(rJ) -SinD(rJ);0 SinD(rJ) CosD(rJ)];
66    mPitch        = [CosD(pK) 0 -SinD(pK);0 1 0;SinD(pK) 0 CosD(pK)];
67    qLVLHToBody   = Mat2Q(mRoll*mPitch);
68    qECIToBody    = QMult(qECIToLVLH(:,k),qLVLHToBody);
```

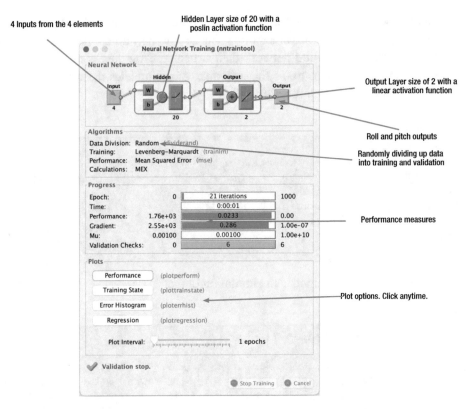

Figure 13.5: *Training window. This window provides information on the training process. We've high-lighted the elements.*

```
69    y           = StaticEarthSensor(qECIToBody,r(:,k),d);
70    c(:,k)      = sim(net,y');
71  end
72
73  [t,tL]  = TimeLabl(t);
74  s           = sprintf('Roll = %8.2f deg Pitch = %8.2f deg',roll,pitch);
75
76  PlotSet(t,c,'x label',tL,'y label', {'Roll' 'Pitch'},'figure title',s)
```

Figure 13.5 shows the feedforward training GUI.

The output of net has all of the parameters used to train the neural network. You can customize everything by editing net after it is created by feedforwardnet.

Figure 13.6 gives the results for the neural network. The image on the upper left is the sensor. The lower left shows the training data. The plots in the upper right show the performance

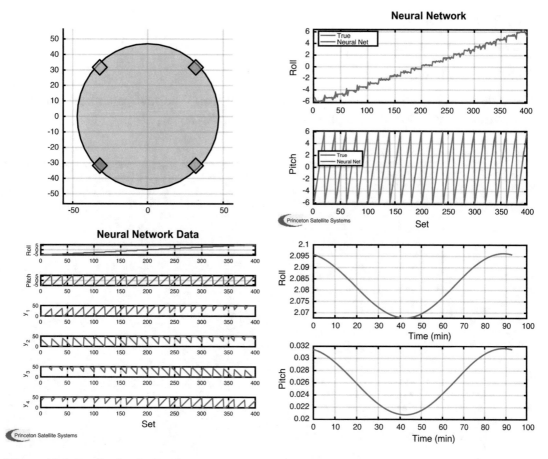

Figure 13.6: *Results for the Earth sensor.*

for all combinations of roll and pitch angles. The pitch measurement is very clean for all the sets. Roll degrades at high roll angles but shows errors even at low roll angles depending on the value of pitch.

13.5   Segmented Sensor Neural Network

13.5.1   Problem

Now we will convert segmented sensor inputs from multiple detectors to roll and pitch measurements.

13.5.2 Solution

Use a feedforward neural network.

13.5.3 How It Works

The code is very similar to that for the previous recipe. The first part of the script, NN SegmentedEarthSensor.m, generates the training data.

NNSegmentedEarthSensor.m

```
1  %% Demonstrate LEO static segmented Earth Sensor using a neural network
2  % The neural network is trained using known roll and pitch.
3
4  degToRad      = pi/180;
5  rE            = Constant('equatorial radius earth');
6  [el,jD0]      = ISSOrbit;
7  d             = SegmentedEarthSensor;
8  [r,v,t]       = RVOrbGen(el);
9  rMean         = mean(Mag(r));
10 qECIToLVLH    = QLVLH(r,v);
11 n             = 20;
12 roll          = linspace(-6,6,n);
13 pitch         = linspace(-6,6,n);
14 i             = 1;
15 y             = zeros(40,n*n);
16 x             = zeros(2,n*n);
17
18 SegmentedEarthSensor(qECIToLVLH(:,1),r(:,1),d)
19
20 for j = 1:n
21   for k = 1:n
22     rJ          = roll(j);
23     pK          = pitch(k);
24     mRoll       = [1 0 0;0 CosD(rJ) -SinD(rJ);0 SinD(rJ) CosD(rJ)];
25     mPitch      = [CosD(pK) 0 -SinD(pK);0 1 0;SinD(pK) 0 CosD(pK)];
26     qLVLHToBody = Mat2Q(mRoll*mPitch);
27     qECIToBody  = QMult(qECIToLVLH(:,1),qLVLHToBody);
28
29     y(:,i)      = SegmentedEarthSensor(qECIToBody,r(:,1),d);
30     x(:,i)      = [roll(j);pitch(k)];
31     i           = i + 1;
32   end
33 end
```

The second part trains the feedforward neural network.

NNSegmentedEarthSensor.m

```
35  % Neural net training data
36
37  net         = feedforwardnet(20);
38
39  net         = configure( net, y, x );
40  net.layers{1}.transferFcn = 'poslin'; % purelin
41  net.name    = 'Earth Sensor';
42  net         = train(net,y,x);
```

The last part tests the neural network.

NNSegmentedEarthSensor.m

```
44  %% Test
45  c           = sim(net,y);
46  leg         = {'True' 'Neural Net'};
47
48  PlotSet(1:size(c,2),[x;c],'x label','Set',...
49      'y label',{'Roll' 'Pitch'},'figure title','Neural Network',...
50      'plot set',{[1 3],[2 4]},'legend',{leg leg});
51
52  yS = zeros(4,size(y,2));
53  for k = 1:4
54      j = 10*k-9:10*k;
55      yS(k,:) = mean(y(j,:));
56  end
57  yL = {'Roll' 'Pitch' 'y_1' 'y_2' 'y_3' 'y_4'};
58  PlotSet(1:size(c,2),[x;yS],'x label','Set','y label',yL,...
59      'figure title','Neural Network Data')
```

Figure 13.7 shows the feedforward training GUI. There are now 40 inputs instead of 4. We still use the same activation function for the hidden layers.

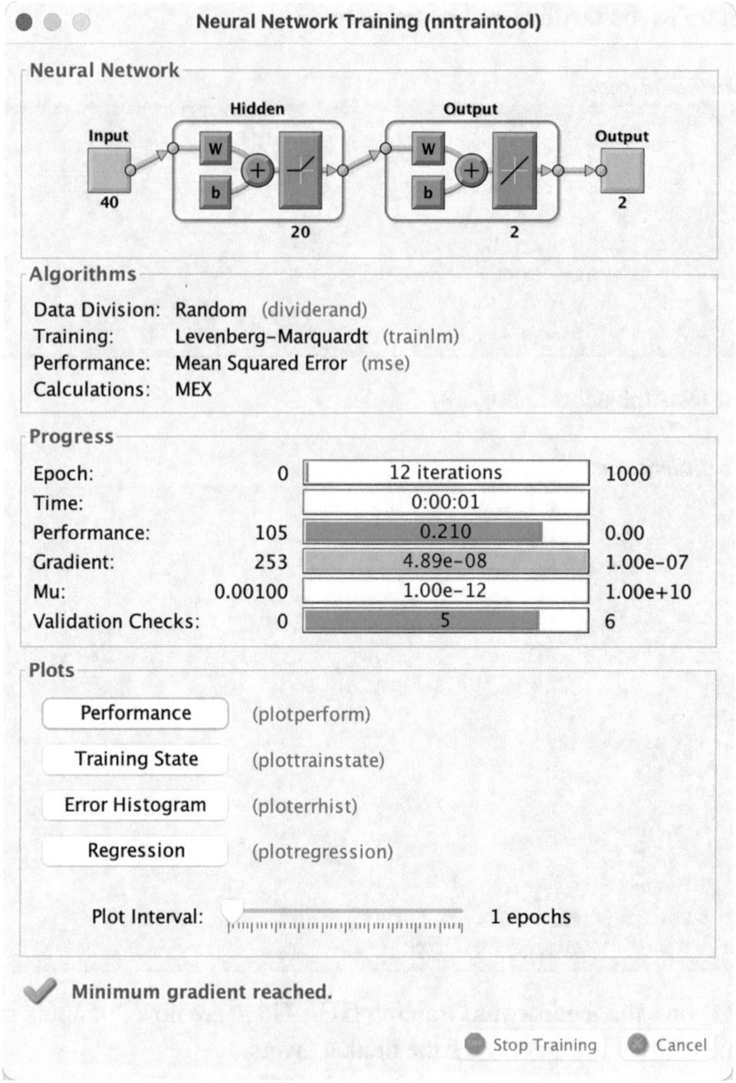

Figure 13.7: *Training window for the segmented sensor.*

Figure 13.8 gives the results for the segmented sensor. The sensor is less accurate than the linear sensor given in the previous example. For the linear sensor, each of the four elements has infinite resolution. In this sensor, each aggregate element only has ten elements to represent the edge of the Earth. Therefore, it only knows its orientation to the angular resolution of one segment. Both pitch and roll show errors over the entire range of angles. Still, the sensor is good to within about a degree.

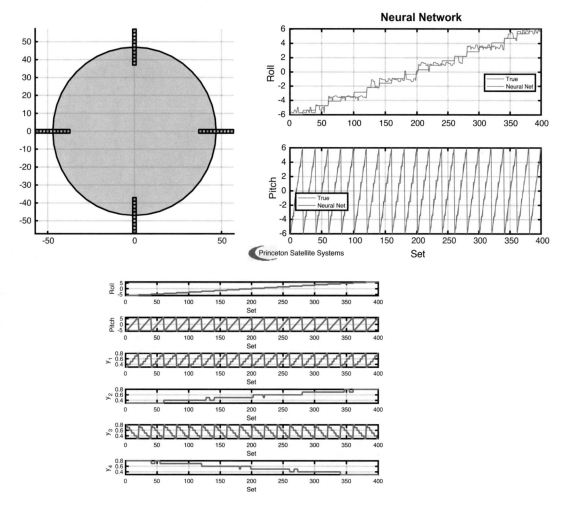

Figure 13.8: *Results for the Earth sensor.*

CHAPTER 14

■ ■ ■

Generative Modeling of Music

14.1 Introduction

Generative machine learning (ML) models are a class of models that allow you to create new data by modeling the data generating distribution. For example, a generative model trained on images of human faces would learn what features constitute a realistic human face and how to combine them to generate novel human face images. For a fun demonstration of the power of ML-based human face generation, check out [44].

This is in contrast to a discriminative model that learns an association between a set of labels and the training inputs. Staying with our face example, a discriminative model might predict the age of a person given an image of their face. In this case, the input is the image of the face, and the label is the numerical age. Labels can also be used in generative models, as we shall highlight shortly in Section 14.2.

Generative models are used in a wide variety of applications from drug design to language models for better chatbots and autocomplete features. Generative models are also used in data augmentation to train better discriminative models, especially in situations where training data is difficult or expensive to obtain. Finally, generative models are widely used by artists and composers to inspire or augment their work. In this chapter, we will investigate generative modeling through this artistic lens by implementing a simple generative model of Bach chorales.

14.2 Generative Modeling Description

We would like to preface these next few sections by stating that what follows is not meant to be mathematically rigorous, even if it uses mathematical terms. Instead, we hope to capture the intuition of generative models. Let us start by assuming that our dataset $\{x_1, x_2, \ldots, x_N\}$ comes from a probability distribution P(X), where X is a random variable. If we had access to this distribution, we could obtain an example x_i by sampling from the distribution ($x_i \sim P(X)$). In general, we do not know this data generating distribution.

In generative modeling, our aim is to approximate this distribution so that we can sample from our approximation and generate new, similar data. In contrast, in discriminative modeling, we try to learn a mapping from the items in the dataset to a set of labels $Y = \{y_1, \ldots, y_M\}$.

© Michael Paluszek, Stephanie Thomas, Eric Ham 2022
M. Paluszek et al., *Practical MATLAB Deep Learning*,
https://doi.org/10.1007/978-1-4842-7912-0_14

In many instances of the generative modeling problem, we are not concerned with labeled datasets. For example, if our aim was to model the distribution that generates faces, we do not explicitly care about features like the age or emotion of the person in a training example. However, in some cases, we might want to train a model that can generate faces that satisfy certain conditions. For example, say we want to be able to tell the model to generate a face that corresponds to someone under the age of 30 or a face that corresponds to someone with an age greater than or equal to 30. We could define X = {human faces of varying ages} and Y = {1 if age < 30, 0 if age >= 30}. We would then task our generative model with learning the joint distribution P(X, Y).

14.3   Problem: Music Generation

The example that we will use to illustrate generative deep learning is music generation. We can think of music as a time series, where at each time point a group of musical features, like pitch and volume, take fixed values from the corresponding sets that contain their possible values. To create a generative model of this, we need to model the distribution that generates this type of data. One way to approach this is to suppose the data is being generated in a sequence, where at time step t, the probability for a particular combination of musical features $[F_1, \ldots, F_p]$ taking values $[f_1, \ldots, f_p]$ is equal to $P(F_1 = f_1, \ldots, F_p = f_p | x_t, x_{t-1}, \ldots x_{t-n})$, where each x_i is a group of feature values for the ith time step and n is the number of previous inputs to be considered by the model when making a prediction. Note that for some models like temporal convolutional networks (TCNs) and transformers (both described in more detail in Section 14.6), $n \in \{1, \ldots, context\_length\}$, where $context\_length$ is a fixed scalar value, whereas in recurrent architectures, implicitly, $n \in \{1, \ldots, \infty\}$. We say implicit here because recurrent models typically only take one time step as input at a time; however, they keep track of previous inputs through their hidden states. In other words, the prediction at time step t depends on these past inputs implicitly, unlike in TCNs and transformers.

From here on, we will refer to x_i as a musical "event" and will note that a musical event is a set of musical feature values just like the model output. We can therefore rewrite the model output as an x_i, and combine it with the previous probability formulation to form the probability distribution over possible subsequent musical events, $P(x_{t+1} | x_t, x_{t-1}, \ldots x_{t-n})$. This distribution tells us that the probability of a particular musical event occurring depends only on the events that precede it. This is a common formulation used in time series modeling as information about events that have not yet occurred is often difficult to come by in practice, and past information is often sufficient for accurate prediction. For example, if you were to train a model to predict tomorrow's weather, using the weather of today and the past week would likely give you a good estimate, and acquiring information on the weather for the day after tomorrow would be impossible.

Now that we have established that our data is generated sequentially according to the distribution $P(x_{t+1} | x_t, x_{t-1}, \ldots x_{t-n})$, we need only to train a model that takes $x_t, x_{t-1}, \ldots x_{t-n}$ as input and outputs $P(x_{t+1})$ to build a generative model of this data.

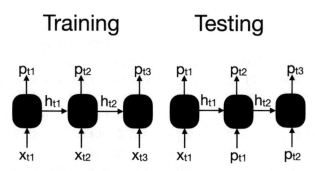

Figure 14.1: *Here is an illustration of the training and testing regimes for a recurrent neural network in our time series modeling problem formulation. On the left, we have an input [x_{t1}, x_{t2}, x_{t3}], which we pass in one at a time. You can imagine that there is a + between each t and the subsequent number (e.g., read t1 as t + 1), but we have omitted these for a cleaner figure. At each time step ti, the model makes a prediction $p_{t(i+1)}$ and updates its hidden state (the horizontal arrows). Upon moving to the next time step during training, the true value that the model was trying to predict is given to the model as input (x_{t2}, followed by x_{t3} in this case). In testing, on the right, however, the model uses its own predictions as inputs after being given x_{t1}. This is the "generative" mode of the model, in which it can continue predicting indefinitely.*

One benefit of these kinds of models is that training them is straightforward. In the typical case, where your goal is just to predict the next time step, your training inputs would be a set of sequences, and your outputs would be these same sequences shifted forward by one. This means that the true output at time t would be x_{t+1}, and at $t + 1$ the true output would be x_{t+2}, and so on. In fact, this is exactly how you would format your data if you were to use a TCN or a transformer as your model. In the case of recurrent architectures like the LSTM, the model only receives one time step at a time, but still "implicitly" utilizes previous inputs as discussed earlier. For these architectures, the output would also be the input shifted forward by one.

A second advantage is that this formulation allows unlimited generation. In "testing," you provide the model with an input, either a musical sequence or some form of "start" token, and the model makes predictions using this input until it reaches the input's end. At this point, the model takes its own predictions as subsequent inputs. As each prediction generates a new input, the model can continue to generate indefinitely. This is visualized in Figure 14.1.

14.4   Solution

We start by simplifying the problem. Music is a very complex domain, with each note carrying a complex set of features including pitch, volume, timbre, onset, duration, and so on. All of these features are captured in raw audio recordings of music. Unfortunately, raw audio can also include extraneous information like nonmusical sounds and noise from a variety of sources. In general, this means that models trained on raw audio require much more data and computational resources to help them learn to distinguish between relevant and irrelevant information in a reasonable timeframe. There are approaches to train models with raw audio to generate music, which we will discuss later in Section 14.6, but due to the challenges associated with using it, we opt to use MIDI data instead. MIDI is a data format intended for electronic music notation and can be used by a variety of music applications like GarageBand. Our MIDI dataset consists of 100 Bach chorales derived from [10]. Each chorale originally had four voices, but only the soprano voice is used here. MIDI data keeps track of a number of musical variables, but for simplicity, we will focus on predicting note pitch and duration. The collection of these variables therefore constitutes our previously defined musical "event."

A prediction of note onset is also possible, but for simplicity, each generated note will be treated as if it were played immediately after the end of the preceding note. We leave it as an exercise to the reader to additionally predict onset to allow for overlapping notes and spaces where no note is played (rests).

For the deep learning (DL) model, we chose a Long Short-Term Memory (LSTM) network with fully connected hidden and output layers. A diagram of an LSTM cell can be seen in Figure 14.2. We chose an LSTM as it is a sequence model and is therefore suited to our formulation of this problem as a time series modeling task. The LSTM is also a well-studied model and is already implemented as a layer in MATLAB, making it more straightforward to implement than other sequence models. Finally, the LSTM has the ability to choose what aspects of its input to remember and what to forget, which in theory allows it to keep track of long-range musical features from chord progressions to melodies.

Like a vanilla RNN, which can also be seen in Figure 14.2, the LSTM has a hidden state, which allows it to propagate information forward for future computation. However, as RNNs struggle with handling tasks where temporally distant information must be recalled, the LSTM augments this short-term memory with the long-term memory of a cell state. The cell state is generated by combining current information from the input with past information from the model's previous hidden and cell states. Including the model's input in the cell state computation allows the model to store information from its inputs, while including the previous states allows past information to be propagated forward. In other words, an LSTM could, in theory, recall its first input, regardless of how many inputs it has seen since.

The main factor that differentiates the cell state from the RNN's hidden state is the LSTM's use of "gates" to determine what information is passed forward to compute the next cell state. The actual gating is carried out by multiplying the gate's input with a set of weights and outputting the result. For example, in the "Forget" gate, values in the past cell state that the gate decides to transmit are multiplied by weights that are close to one, whereas values that it decides to forget are multiplied by weights that are close to zero. These weights are generated by multiplying what we call a "gate keeper" vector with the gate's weight matrix and passing this output through an activation function that maps it to a value within a particular range. One of the most common choices for this activation function, and the one used in the "Forget" gate, is the sigmoid, which will map the output to a value between zero and one. In the case of the "Forget" gate, the gate keeper is the concatenation of the input with the past hidden state. The gate's weight matrix is learned during model training.

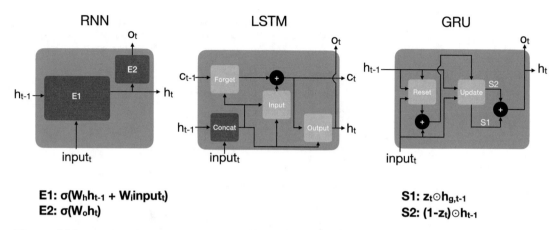

E1: σ(W$_h$h$_{t-1}$ + W$_i$input$_t$)
E2: σ(W$_o$h$_t$)

S1: z$_t$⊙h$_{g,t-1}$
S2: (1-z$_t$)⊙h$_{t-1}$

Figure 14.2: *LSTM and other recurrent architectures. Here, we show diagrams of RNN, LSTM, and GRU cells. Gates are visualized in green. For consistency, the gate inputs line up with the gate outputs, whereas the gate keepers are perpendicular to the gate inputs. The GRU update gate is a notable exception. In this case, the factor generated by the gate, z_t, is applied to $h_{g,t-1}$, a combination of the input and the output of the reset gate (a gated version of the previous hidden state), and $(1-z_t)$ is applied to the previous hidden state (h_{t-1}). The application of these factors is through the Hadamard product, as seen in the figure. Effectively, both $h_{g,t-1}$ and h_{t-1} are being "gated," which leads to the difference in the diagram. As an aside, note that z_t linearly parameterizes the degree to which the past hidden state or the past hidden state/new input combination is represented in the output. If $z_t = 0, o_t = h_{t-1}$, if $z_t = 1, o_t = h_{g,t-1}$, and if $0 < z_t < 1, o_t$ is a linear combination of these values. Also note that for both the LSTM and GRU, the next hidden state is also the output. Finally, note that we grouped model components into gates in such a way as to minimize the complexity of our diagrams and omitted bias vectors and some weight matrices for a similar effect. Some accuracy has therefore been sacrificed to help foster an intuitive understanding of the models. This figure was inspired by a figure in [25] and benefited from the information in [30].*

Figure 14.3 illustrates a general gating mechanism as seen in LSTMs. This gating mechanism improves the performance of LSTMs on tasks that require recollection of specific past information and make them less susceptible to the vanishing gradient problem of vanilla recurrent neural networks (RNNs), where the products of gradients become so small that the output neurons lose the ability to update their weights. Note that although in theory, an LSTM seems quite powerful, in practice, the model only "remembers" what it is trained to remember. If the model is not shown the utility of recalling past inputs, it will not do so. Common methods of highlighting this utility are constructing a training set where success is only possible if long-term trends are attended to, or by penalizing the model for not doing so in the loss function. For the sake of this tutorial, the internal workings of the LSTM are not that important, and we encourage the reader to read about LSTMs and other RNNs if their inner workings and theoretical guarantees are of interest.

Gating Mechanism

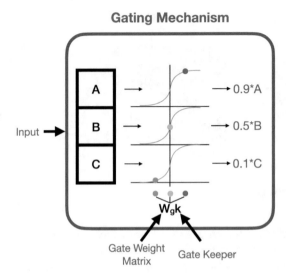

Figure 14.3: *LSTM gating intuition with a sigmoid activation. Here, we see a three-dimensional input to the left and the weight matrix and gate keeper at the bottom. The product of the weight matrix and gate keeper produce a value for each input (the colored dots), which when passed through the sigmoid activation function, generate weights between zero and one. Note that by using the sigmoid activation (as opposed to the softmax, for example), these factors need not add up to one. This means that whether or not the gate transmits or blocks one component of the input is not constrained by the decision made for another component of the input. Finally, these factors are used to scale the input to produce the output of the gate. As one specific example of an LSTM gate, take the "Forget gate" as seen in Figure 14.2. In the forget gate, the input is the previous cell state, and the gate keeper is the concatenation of the previous hidden state with the current input. Intuitively, this means that we use the most current information from the input, with short-term memory of past states and inputs, to determine which aspects of the cell state should be passed forward and which aspects should be forgotten.*

14.5 Implementation

Before diving into the implementation of our solution to this problem, we note that this chapter requires some additional toolboxes:

1. Statistics and Machine Learning Toolbox by MathWorks

2. MIDI Toolbox [43]

Once we have those toolboxes downloaded, we are ready to start coding. As was noted previously, MATLAB already implements an LSTM with its lstmLayer. In order to utilize this layer, the model must be built with a sequential input layer. This model formulation allows you to pass sequences of inputs to the model and have it process them in the proper temporal order. With a batch size of one, there is no need to pad the inputs; however, should you wish to have the model observe more sequences before an update, padding or cropping will be required as MATLAB requires that all sequences in a single batch be the same length.

Another implementation detail is whether to formulate the problem as classification or regression. In a regression formulation, the fully connected layer at the output would have two outputs, one for duration and one for pitch. The outputs would be continuous, allowing the model to generalize to unseen pitches and note durations. One major problem with using regression for generative modeling is that you lose the ability to manually alter the way you sample from your model's output distribution. As we shall see at the end of this section and the beginning of Section 14.6, some sampling procedures perform much better than others, and having this degree of freedom in choosing the way you sample gives you the ability to greatly improve the quality of the music your model generates. For this reason, we choose to formulate our problem as classification.

In classification, the output space is divided into a set of classes that the model has to assign probabilities to. A possible formulation of such a problem would be a multi-output model where one output corresponds to pitch and the other corresponds to duration. This would allow shared processing of the model input, but separate pitch-specific and duration-specific processing before the model output. This mode of operation is more difficult to implement in MATLAB as it requires a custom training loop. Our chosen approach is to instead have the model predict the joint distribution over pitch and duration as it allows the model to predict a single vector output. We will label this model as the "single-output" model to contrast it with the "multi-output" model we defined before. We discuss both methods here to highlight their differences.

The principal difference between these two models is how they model the distribution of music. As stated in the introduction to generative modeling, our goal here is to model the distribution that generates the data. In this case, this is the distribution that generates the soprano voice of Bach chorales. Directly modeling the distribution allows us to sample from it to generate new examples. In the multi-output model, the model generates a distribution over pitches and a distribution over durations, which we would sample separately. In the single-output model, we sample from the joint distribution over pitch and duration.

The joint distribution inherently contains more information than the marginal distributions, which are modeled by the multi-output model, except in the case that the distributions are independent. There is likely some dependence between pitch and duration, though we are not convinced that the dependence is significant enough to choose the single-output model over the multi-output model for this reason.

However, a multi-output model has a significant benefit, which is that the size of the output space is much smaller, and separate processing can be applied to each output. This characteristic becomes much more important as we increase the number of outputs. In our simplified dataset, there are 8 possible durations and 20 possible pitches, making the output space of size 160. If there were a third variable, say of size 10, we would now have an output of size 1600. Larger output spaces in classification problems invariably lead to many low probability outputs and, thus, small gradients. This can contribute to the vanishing gradient problem and can make learning hard without a large dataset and an abundance of computational resources.

To contrast, in this scenario, the multi-output model would have an output of size 20 + 8 + 10 = 38, which is much more manageable. However, as mentioned previously, we implement the single-output model here as it is still effective in our simplified problem and is easier to implement in MATLAB.

Finally, we would like to note that the output size is inherently arbitrary. Duration and pitch are both actually continuous variables that we have discretized in order to format our problem as classification. It should be noted that these assumptions are not unreasonable when handling classical music. In classical music, there is a fixed set of pitches that can be realized (A, Bb, C, ...), and durations generally correspond to a fixed set $= \{16th, 8th, quarter \ldots\}$. There are of course exceptions to these rules, and we leave the construction of a model that accounts for these exceptions as an exercise to the reader. We further restrict the possible pitches and durations to those found in the training set. However, we do not restrict the possible *combinations* of pitches and durations to those in the training set, as can be seen in line 53 of the code.

The code that follows comes from Music_LSTM.m. It constructs and trains the single-output LSTM for music generation. It then generates music using training set examples as starting seeds.

In this section of the code, we load and preprocess our dataset.

```
1   % clear workspace
2   clear
3
4   % load data
5   % format: [st, pitch, dur, ks, ts, f]
6   % st = start time (ignored)
7   % dur = duration
8   % ks = key signature (ignored)
9   % ts = time signature (ignored)
10  % f = fermata (ignored)
11  mats = load('chorales_testmat.mat');
12  fn = fieldnames(mats);
13
14  % store our data (pitches and durations) in a cell array.
15  chorale_cell = cell(numel(fn),1);
16  for k=1:numel(fn)
17      chorale_cell{k} = mats.(fn{k})(:,2:3);  % get pitch + duration
18  end
19
20  % get set of unique pitches and set of unique durations
21  big_seq = cat(1, chorale_cell{:});
22  pitch_labels = sort(unique(big_seq(:,1)));
23  dur_labels = sort(unique(big_seq(:,2)));
24  dl = length(dur_labels);
25  np = length(pitch_labels);
26
27  % get indices corresponding to each musical event (pitch, duration)'s
28  % location in the label arrays.
29  % Use these indices along with encode_pd to generate a single label for
30  % each event
31  % label cell contains these "encoded" labels
32  label_cell = cell(numel(fn, 1));
```

```
33  for i=1:numel(chorale_cell)
34      seq = chorale_cell{i};
35      ls = length(seq);
36      chorale_labels = zeros(ls,1);
37      for j = 1:ls
38          pi = find(pitch_labels == seq(j,1));
39          di = find(dur_labels == seq(j, 2));
40          chorale_labels(j,1) = encode_pd(pi, di, dl);
41      end
42      label_cell{i} = chorale_labels;
43  end
44
45  % divide data into input (X) and output (Y), where Y is just
46  % X shifted by 1 to the future.
47  [X, Y] = get_xy(label_cell);
48
49  % convert labels to categorical data for classification
50  % note that we have a class for each possible combination of pitches
51  % and durations (20*8 = 160), intead of just for the combinations that
       appear
52  % in the dataset. This may allow for more interesting pieces that
       contain
53  % patterns not seen in the training set.
54  Y = cellfun(@(x) categorical(x, 1:np*dl) ,Y,'UniformOutput',false);
```

In this code block, we define the LSTM, set training parameters, and train the LSTM.

```
57  % define LSTM
58  % adapted from https://www.mathworks.com/help/deeplearning/ug/time-
       series-forecasting-using-deep-learning.html
59  % define input, layer and output sizes
60  num_hidden_units = 128;
61  input_shape = 1; % num features = 1, since we have combined our 2
       features
62  num_outputs = np*dl;
63
64  % Define model layers
65  layers = [ ...
66      sequenceInputLayer(input_shape, Name='seq in')
67      lstmLayer(num_hidden_units,'OutputMode','sequence', Name='lstm')
68      softmaxLayer
69      fullyConnectedLayer(num_outputs, Name='fc1')
70      softmaxLayer
71      fullyConnectedLayer(num_outputs, Name='fc2')
72      softmaxLayer
73      classificationLayer(Name='co')];
74
75  % optional L2 regularization
76  layers(4) = setL2Factor(layers(4),'Weights',2);
77  layers(6) = setL2Factor(layers(6),'Weights',2);
78
```

```
79  % Visualize the network
80  analyzeNetwork(layers)
81
82  % define training parameters
83  max_epochs = 400;
84  % use mini batch size of 1 to avoid padding
85  mini_bs = 1;
86  options = trainingOptions('adam', ...
87      'MaxEpochs',max_epochs, ...
88      'MiniBatchSize', mini_bs, ...
89      'InitialLearnRate',0.01, ...
90      'LearnRateSchedule','piecewise', ...
91      'Shuffle', 'every-epoch',...
92      'Verbose',0, ...
93      'LearnRateDropPeriod',100, ...
94      'LearnRateDropFactor',0.5, ...
95      'Plots','training-progress');
96
97  % train LSTM
98  net = trainNetwork(X,Y,layers,options);
99  disp('Training Complete')
```

In this code block, we use the LSTM and regular sampling to generate music using parts of different training examples as seeds.

```
102 generate(net, X{50}, pitch_labels, dur_labels, 'new_test50_simple', 'na
        ', 2);
103 generate(net, X{25}, pitch_labels, dur_labels, 'new_test25_simple', 'na
        ', 2);
104 generate(net, X{10}, pitch_labels, dur_labels, 'new_test10_simple', 'na
        ', 2);
```

The following functions are called during the execution of the preceding script. The first function uses the trained LSTM to generate new music. The subsequent functions are used for preprocessing and saving outputs.

Function for using trained LSTM to generate new music

```
106 % Uses trained LSTM to generate new music. Saves outputs as MIDI files
107 % to be played and or visualized in an application that supports MIDI
108 % net: trained LSTM
109 % test_seq: sequence from the training set to use as seed
110 % pitches: possible pitches
111 % durs: possible durations
112 % name: output name
113 % samp mode: 'greedy' for greedy sampling. else: regular.
114 % gen_amt: sets the amount to generate as a function of the input size.
115 % outputs: are described in the within function comments
116 function generate(net, test_seq, pitches, durs, name, samp_mode,
        gen_amt)
```

278

```
117    test1 = test_seq;
118    % seed size
119    hs = ceil(length(test1)/2);
120    % set amount you want to generate
121    num_test = gen_amt*hs;
122    for i = 1:num_test
123        if i == 1
124            % use first half of song for the first prediction
125            history = test1(:,1:hs);
126            % to see what happens if only first event of seed provided
127            %history = test1(:,1);
128        else
129            % concatenate past history with last prediction to continue
130            % generating
131            history = [history, last_pred];
132        end
133        % make prediction and update hidden state
134        [net,pred] = predictAndUpdateState(net,history,'
               ExecutionEnvironment','cpu');
135        last_preds = pred(:, end); % get last prediction for
               concatenation
136        % In greedy mode, you take the highest probability event
137        if strcmp(samp_mode, 'greedy')
138            [max_v, idx] = max(last_preds); % greedy
139            last_pred = idx;
140        else
141            % in sampling mode, you sample according to the
                 distribution given by the model
142            last_pred = randsample(length(last_preds),1,true,last_preds
                 );
143        end
144
145    end
146    % concatenate last prediction
147    history = [history, last_pred];
148
149    % save the entire training example
150    [ps,ds] = decode_label(test1, pitches, durs);
151    save_as_midi(ps, ds, strcat(name, '_samp_class_full_test'));
152    % save the first half of the example to see where generation starts
153    [ps,ds] = decode_label(test1(:,1:hs), pitches, durs);
154    save_as_midi(ps, ds, strcat(name,'_samp_class_in'));
155    %save the input plus the generated output.
156    [ps,ds] = decode_label(history, pitches, durs);
157    save_as_midi(ps, ds, strcat(name, '_samp_class_in_plus_gen'));
158 end
```

279

Data preprocessing functions

```
160   % converts each sequence to input (x) and output (y) for our next step
161   % prediction objective. output is input shifted forward by 1.
162   function [x, y]= get_xy(seqs)
163       x = cell(length(seqs),1);
164       y = cell(length(seqs),1);
165       for i = 1:length(seqs)
166           x{i} = seqs{i}(1:end-1,:);
167           y{i} = seqs{i}(2:end,:);
168       end
169       x = cellfun(@transpose,x,'UniformOutput',false);
170       y = cellfun(@transpose,y,'UniformOutput',false);
171   end
172
173   % creates a label for a given pitch and duration combination
174   % i = pitch index --> i = 1:length(pitch_labels)
175   % j = duration index --> j = 1:length(dur_labels)
176   % dl = number of possible durations = length(dur_labels)
177   % label for i,j combination = (i-1)*length(dur_labels) + j
178   function label = encode_pd(i,j, dl)
179       label = (i-1)*dl + j;
180   end
181
182   % reverses encode_pd to get pitch and duration from the label
183   % note that label is a vector here (from the model output)
184   % pitches = possible pitches
185   % durations = possible durations
186   function [pitch,duration] = decode_label(label, pitches, durations)
187       dl = length(durations);
188       j = mod(label, dl);
189       % if j = dl, label = i*dl, so label%dl = 0.
190       % For our math to work out, we want j = dl in this case, not 0.
191       % so set all entries p where the j(p) = 0 to dl.
192       % since matlab indexing starts at 1, we don't have to worry about
193       % a case where j actually equals 0.
194       j(find(j == 0)) = dl;
195       i = (label - j + dl)/dl;
196       pitch = pitches(i);
197       duration = durations(j);
198   end
```

Function for saving generated music to be later visualized or played in an application like Apple's GarageBand

```
200   % converts generated output to nmat and saves as midi
201   % uses nmat functionality from MIDI Toolbox:
202   % https://www.jyu.fi/hytk/fi/laitokset/mutku/en/research/materials/
          miditoolbox
203   %[:, st(b) dur(b) channel pitch velocity st(s) dur(s)] <-- MIDI Toolbox
          nmat format
```

```
204   % b: beats, s: seconds
205   % pitches and durs (durations) generated by our model
206   % name: output file name
207   function save_as_midi(pitches, durs, name)
208       out_s = length(pitches);
209       nmat = zeros(out_s, 7);
210       nmat(:,2) = durs; % duration
211       nmat(:, 4) = pitches; % pitch
212
213       onsets = zeros(out_s,1);
214       onsets(1) = 0; % onsets start at 0 in nmat format
215       for i = 1:out_s-1
216           % get current onset by adding previous duration to previous
                  onset
217           onsets(i+1) = onsets(i) + durs(i);
218       end
219       nmat(:,1) = onsets; % onsets
220       % channel range: 1-16, choice is arbitrary here with one channel
221       nmat(:,3) = ones(out_s,1);
222       velocity = 50; %  1-127
223       % velocity (volume), choice is somewhat arbitrary here
224       nmat(:,5) = velocity*ones(out_s,1);
225       % these determine tempo
226       t_ratio = 15;
227       nmat(:,6) = nmat(:,1)/t_ratio; % start time (s)
228       nmat(:,7) = nmat(:,2)/t_ratio; % duration (s)
229       writemidi(nmat, strcat(name,'.midi'));
230   end
```

The outputs can be played in any MIDI-compatible application. Figure 14.4 shows the visualization of one model output in GarageBand.

Here, we will briefly discuss the results we achieved with this model so that the reader will be better equipped to replicate them. You can see our loss and accuracy curves in Figure 14.5. We note that the model appears to learn most of what it needs within the first 50 or so epochs.

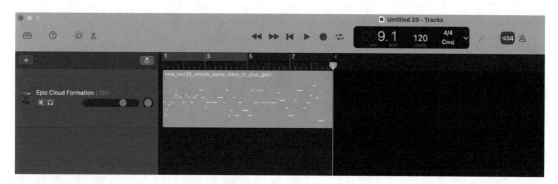

Figure 14.4: *The model-generated MIDI file new_test25_simple_samp_class_in_plus_gen.midi in the GarageBand application.*

We also note that the average accuracy value appears to approach 20% by the end of training, which is much greater than chance (1/160 = .625%).

Once the trained model is generated, there are several ways to use it to produce music. Two of these are contained in our generate function. The first is greedy sampling, in which we take the (pitch, duration) combination with the highest probability (as assigned by the model) as the next musical event. In the second, we sample from the distribution produced by the model. This means that we take the event corresponding to index i in the output vector with probability $P(i)$, as assigned by the model. As you can see in Figure 14.6, the latter method produces superior results. In our generate function, we first provide the model with part of a training example to get it started before switching to feeding it its own predictions as inputs. This gives the model an opportunity to construct hidden and cell states based on the higher-level features of the piece, like mood or style. If we had trained on composers other than Bach, another high-level feature would be the composer. By providing a short sample of a piece as a seed, we encourage the model to generate music that fits the style, mood, and so on of its seed. In Section 14.6, we discuss probing the feature space of variational autoencoders to control what the model generates, and this is one way to achieve a similar effect with LSTMs.

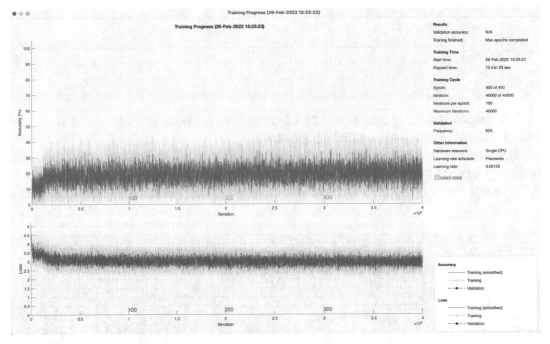

Figure 14.5: *Model training curves. Accuracy quickly jumps from around 0% to around 10% within the first epoch. By epoch 15, the accuracy is up to around 20% and then hovers around there for the remainder of training. This indicates training could likely have been stopped much sooner without sacrificing model performance. Also note that the x-axis of these plots is in "iterations," but as can be seen on the right of the figure, there are 100 iterations per epoch here. The epoch markers can also be seen above the iterations in steps of 100.*

You may be wondering why we have not computed the validation loss for our model. There is nothing wrong with doing this; however, by comparing the true output and the generated output in Figure 14.6, we can clearly see that the model is not overfitting to the training set.

14.6 Alternative Methods

The LSTM that we have constructed can definitely act as a generative model of music as we can sample from its output distribution to produce new, reasonably convincing series of musical events. That being said, it can definitely be improved. One way to improve the model's output without changing the model itself is to add a temperature parameter to the sampling procedure [40]. To do this, we pass the probabilities the model assigns to each potential output through what's called the "freezing function" as seen in Equation 14.1. In Table 14.1, you can see how different temperatures affect a simple probability distribution, but overall, as $t \to 0$, the probabilities become more skewed, with the maximum probability approaching 1 and the others approaching 0. When you sample from this new distribution, you are effectively taking the

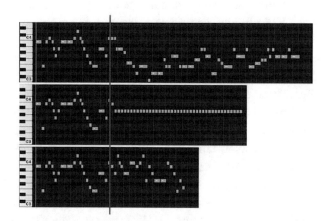

Figure 14.6: *Three versions of the same chorale (tenth in our training set). The top two are generated by our model. The bottom one is the true version of the song. The red line indicates where the model started generating. The top one was generated by sampling from the model's output distribution. The middle one shows what happens when we sample greedily from this distribution (only taking the highest probability note and duration combination). Not only does the sampled version look better, but it sounds much more likely to have come from the true piece.*

Table 14.1: *Effect of temperature parameter on model output probabilities*

Identifier	Input Probabilities	Temperature (τ)	Output Probabilities
Model Output	(0.2, 0.8)	1	(0.2, 0.8)
More Uniform	(0.2, 0.8)	10	(0.4654, 0.5346)
Near Uniform	(0.2, 0.8)	1000	(0.4997, 0.5003)
More Greedy	(0.2, 0.8)	0.5	(0.0588, 0.9412)
Near Greedy	(0.2, 0.8)	0.1	(0.0000, 1.0000)

argmax or using "greedy" sampling. As $t \to \infty$, the new distribution approaches the uniform distribution, with each possible output getting an equal probability. At t=1, you sample from the model according to the probabilities it assigns. In our code, we effectively implemented sampling with $t = 0$ (greedy) and $t = 1$. However, by explicitly using temperature to adjust the output distribution, you can try to find a balance between sampling from the model's output distribution and using random and greedy sampling to get the most realistic sounding music.

$$p_{j,new} = f_{freeze}(p)_j = \frac{p_j^{1/\tau}}{\sum_i p_i^{1/\tau}} \tag{14.1}$$

Another problem with our model is that greedy sampling results in a completely unrealistic piece of generated music. It seems that the model assigns highest probability to what it considers to be a "safe" choice of replicating the same pitch and duration. This tells us that the model is not learning how to generate music as well as we would hope. There are various ways that you could try to improve this, including using a larger dataset and training for longer with more computational resources. You could also change the model hyperparameters like the number of layers or the size of the LSTM hidden states or the training hyperparameters like the learning rate.

In addition to improving the LSTM, you could try to obtain better performance on this task by implementing some alternative ML models. There are several models that could be swapped in for the LSTM easily. These include vanilla RNNs and gated recurrent unit (GRU) RNNs, which can be seen in Figure 14.2. LSTM and GRU models are considered to be superior to vanilla RNNs. Both LSTMs and GRUs use gating to control what information is utilized and stored by the network; however, GRUs are slightly simpler, allowing them to be trained more efficiently with comparable performance. As can be seen in Figure 14.2, one source of this simplicity is combining the hidden and cell states of the LSTM into one hidden state in the GRU. This increase in training efficiency may make a GRU an improvement as it could allow us to reap the benefits of using a larger model (larger models can identify more nuanced patterns) or training on more data (more examples to discover patterns from) with a relatively small increase in training time.

Other models used for time series modeling, but which may require more work to implement, include temporal convolutional neural networks (TCNs) and transformers. Unlike recurrent models, these models do not require computations on previous inputs to be completed before moving to future inputs. Instead, they are given all the inputs within a given input size, also known as context length, at once and generate predictions for each time point in the input in parallel. This allows these networks to be trained in a much more efficient, parallel manner.

TCNs are based on one-dimensional convolutions, which are convolutions where the convolutional kernel is shifted along the sequential dimension of a sequence input. TCNs also use skip connections and dilated causal convolutions. Skip connections allow information to be passed between layers at different depths, and dilated convolutions increase the model's receptive field so that the model can use information farther in the past for its predictions. Finally, causal convolutions are convolutions where any inputs that occur after the current time step are ignored. Wavenet is an example of a model that uses TCN-like characteristics for music generation [28].

Transformers use attention as their fundamental building block. The mechanism of attention is beyond the scope of this text, but you can think of it as a method for pairwise comparison. Though transformers are most often used in language modeling and language comprehension tasks, there is no reason they cannot be used for generative music modeling [16]. As was the case with the GRU, the ability to train TCNs and transformers much more efficiently can make it more feasible to train larger more expressive versions of these models on larger datasets. In addition, by giving the model the entire input on which it is to make predictions, the model is able to access whatever past inputs are relevant for the predictions at each time step and combine information across multiple time scales into complex hierarchical representations. Recurrent models, like LSTMs, on the other hand, only have access to what they decided was worth remembering at some past time step, and so the important information for the current prediction may have been forgotten. That being said, models like TCNs and transformers suffer from the need to define a fixed input size, which limits their ability to operate on past inputs, whereas in theory, an LSTM could use information from its very first input for any future predictions.

Finally, we would like to briefly discuss variational autoencoders (VAEs). These models are too complex to be covered in depth, but the general idea is that like a traditional autoencoder, they map a higher dimensional input to a lower dimensional latent space and back to the higher dimensional input space. Effectively, a VAE learns which features in an input signal are truly necessary to reconstruct the input. What distinguishes the VAE from a traditional autoencoder is that its latent space is parameterized by a multivariate Gaussian distribution. In other words, the VAE approximates the data generating distribution as a multivariate Gaussian, and by sampling from this distribution, new data points can be generated.

When trained properly, the latent space is considered to be "structured," a property that can prove advantageous for generative modeling. This structure corresponds to the mapping of inputs that are more similar to points that are closer together in the latent space and can lead to the presence of "feature axes," or lines that correspond to the variations in a particular feature. This is more clearly illustrated with our face example from Sections 14.1 and 14.2, so we will first consider a VAE trained to model the distribution of human faces. The human face has a number of distinctive features, including lips, eyes, the relative positions of these and other features, and so on. As a simple example, the shape of the lips might constitute a feature axis. Say you locate this axis, take a point on it, and decode it through the VAE decoder. The output is a face with a neutral lip shape. You move to one side of this point on the feature axis, select a new point, and decode it, revealing a smiling face. You take a step further in this direction, and the smile is almost comically large. Now you decide to take a couple steps past your original point in the opposite direction and find that the decoded face is frowning. A key piece of information to take note of is that in an ideally structured latent space, the only part of the image that should change while you traverse the lip shape axis is the shape of the lips. In other words, the different feature axes should be orthogonal in the latent space, such that the features they describe are independent. In simpler terms, you should be able to change one feature without changing another. This is an extremely advantageous property in generative modeling as it gives you substantial control over what you are generating. Once you identify a set of feature axes by encoding faces that differ in specific ways and comparing their latent representations, you can take a given face and perturb it in any number of ways, without resulting in unwanted changes. We give some intuition for a 2D latent space in Figure 14.7. We also should note that to be able to map N features to orthogonal feature axes in the latent space, we will need, at a minimum, an N-dimensional latent space.

Returning to our musical example, should we train a VAE to map a set of musical pieces to a latent space, we might find feature axes for the mood of the piece or its style. We could potentially take a piece by Mozart, encode it, and then slide the encoded point along its style axis to make it sound like jazz or a modern pop song. As another example, we might be able to take a melancholy piece and give it a cheerful spin by moving our latent point along the mood axis.

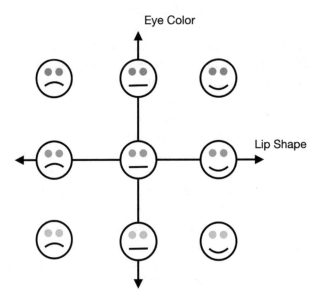

Figure 14.7: *Here, we see a well-structured two-dimensional latent space for faces. The two features pictured are eye color and lip shape. Moving along the lip shape axis from left to right corresponds to changing the lip shape from a frown to neutral to a smile. Traversing the eye color axis from top to bottom corresponds to transitioning from orange eyes to blue eyes and then green eyes. The features are on orthogonal axes so that changing one does not change the other. By moving along both axes, we can generate faces that combine the features. For example, when we move from the center face up and right, we move to a face with orange eyes and a smile. Note that in reality, the faces themselves do not exist in the latent space. The trained encoder and decoder assign meaning to the 2D latent points by creating mappings between them and the faces in the input space.*

As we discussed in Section 14.5, we do have an ability to alter what is generated by our LSTM by seeding its generation with songs from different composers, or of different styles or moods, and so on. However, it would be much easier to isolate particular features of a musical piece and to change them independently with a VAE. One problem with using a VAE by itself to generate music is that the output size of the model is fixed. To generate an arbitrarily long sequence of music, one would need to repeatedly generate musical sequences of a fixed length and string them together. Another problem is that VAEs can be very difficult to train in part because you are balancing the KL loss which tries to push the encodings of all inputs together (toward the normal distribution) and the reconstruction loss which tries to push them apart so that they can be differentiated more easily by the decoder. These losses can be seen in Figure 14.8.

Another way in which VAEs are used for music generation is as a method for handling raw data. As we mentioned previously, raw data is challenging to use as it contains many extraneous features that might distract our model from the music it is supposed to be modeling. Since VAEs can be used to extract the important features in a signal, we can train a VAE to extract musically

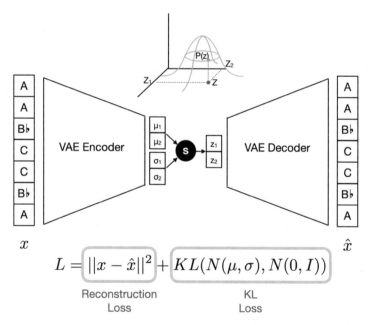

Figure 14.8: *Example VAE training using pitch as input. In this example, we model the latent space as a bivariate normal distribution, so the encoder generates a 2D mean vector and 2D standard deviation vector. In training, we sample from $N(\mu, \sigma)$ to get z. This is signified by the black circle with the white "s" in the figure. In testing, we use μ for z. The decoder then reconstructs the original pitch sequence from z. In training, there is a reconstruction loss that penalizes large differences between the input sequence and the output sequence, and a KL loss, which penalizes the distance of $N(\mu, \sigma)$ from $N(0, I)$, where 0 is the zero vector and I is the identity matrix. This latter loss is based on KL divergence, and it encourages the construction of a structured latent space. The complete loss is the summation of these two losses and can be seen at the bottom of the figure. This figure was inspired by a figure in [33].*

relevant features from raw audio. We can then use one of the previously described sequence models, like an LSTM, to predict future musical events. This technique is a simplification of what the authors of Jukebox did [9].

Hopefully, this chapter gave you some foundational understanding of ML-based generative modeling, especially as it applies to music. There is certainly room for exploration, and we hope that this last section has given you a taste of what is possible in this exciting area.

CHAPTER 15

■ ■ ■

Reinforcement Learning

15.1   Introduction

Reinforcement learning is a machine learning approach in which an intelligent agent learns to take actions to maximize a reward. We will apply this to the design of a Titan landing control system. Reinforcement learning is a tool to approximate solutions that could have been obtained by dynamic programming, but whose exact solutions are computationally intractable [3].

We'll start the chapter by modeling the point-mass dynamics for the vehicle. We will then build a simulation. We'll test the simulation with the vehicle that starts in orbit around Titan, a moon of Jupiter, and with the vehicle in level flight. We will then use optimization to find a trajectory. Finally, we'll design a reinforcement learning algorithm to find a trajectory.

15.2   Titan Lander

The problem, for which we will use reinforcement learning, is a powered landing on Titan. Titan is the second-largest moon in the solar system and is larger than the planet Mercury. Titan has a thick atmosphere which makes it feasible to fly an airplane in the atmosphere using aerodynamic forces for lift. Figure 15.1 shows an image of Titan and its atmospheric density.

We will use a model similar to that in Chapter 9, terrain-based navigation, except that we will simplify it to the planar case. In addition, we will assume that the mass of the vehicle does not vary. Our propulsion system uses a nuclear fusion–powered turboramjet that consumes very little fuel. It heats the incoming atmosphere with the energy from the fusion reaction. The point-mass vehicle equations of motion are different from Chapter 9 in that we consider that the moon is round and that gravity varies in strength as the vehicle descends. The equations are written in cylindrical coordinates:

$$\dot{r} = u_r \tag{15.1}$$

$$\dot{u}_r = \frac{v^2}{r} - \frac{\mu}{r^2} + a_r \tag{15.2}$$

$$\dot{u}_t = -\frac{u_t u_r}{r} + a_t \tag{15.3}$$

where r is the radial position, u_r is the radial velocity, u_t is the tangential velocity, μ is the gravitational parameter for Titan, and a_r and a_t are the accelerations in those directions. We

© Michael Paluszek, Stephanie Thomas, Eric Ham 2022   289
M. Paluszek et al., *Practical MATLAB Deep Learning*,
https://doi.org/10.1007/978-1-4842-7912-0_15

Figure 15.1: *The composite infrared image of Titan is courtesy of NASA.*

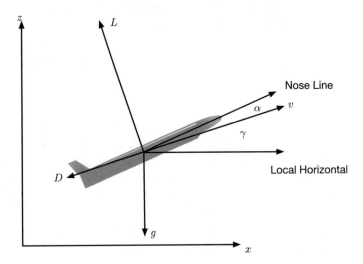

Figure 15.2: *Aircraft model showing lift, drag, and gravity.*

don't integrate the tangential position or angle since our problem will only involve landing. If you wanted to land at a particular spot, after planning your trajectory, you would just pick an orbital position to begin descent so that you end at the right spot. In the diagram in Figure 15.2, the velocity magnitude v is

$$v = \sqrt{u_r^2 + u_t^2} \tag{15.4}$$

The flight path angle γ is not used in this formulation. The gravitational acceleration appears in Equation 15.1.

$$g = \frac{\mu}{r^2} \tag{15.5}$$

We use a very simple aerodynamic model. The lift coefficient c_L is defined as

$$c_L = c_{L_\alpha} \alpha \tag{15.6}$$

The lift coefficient is in reality a nonlinear function of the angle of attack α. It has a maximum angle of attack above which the wing stalls and all lift is lost. This coefficient will also vary with flight speed. For our simple model, we will assume a flat plate with $c_{L_\alpha} = 2\pi$. The drag coefficient is

$$c_D = c_{D_0} + k c_L^2 \tag{15.7}$$

where k is

$$k = \frac{1}{\pi A_R \epsilon} \tag{15.8}$$

A_R is the aspect ratio and ϵ is the Oswald efficiency factor which is typically from 0.8 to 0.95. The efficiency factor is how efficiently lift is coupled to drag. The aspect ratio is the ratio of the wingspan (from the point nearest the fuselage to the tip) and the chord (the length from the front to the back of the wing). This equation is known as the drag polar. When there is no lift, the drag coefficient is c_{D_0}. c_D varies as the square of the lift coefficient. For hypersonic flow, above Mach 4, the following model, based on Newtonian Impact Theory, is

$$L = \rho S V^2 \sin^2 \alpha \cos \alpha \tag{15.9}$$

$$D = \rho S V^2 \left(\sin^3 \alpha + \frac{0.075}{R_E^{1/5} \sqrt{M}} + C_{D_\pi} \right)$$

where C_{D_π} is approximately 0.001. This would need to be meshed with the previous model for when the Mach number drops below four. For the purpose of this chapter, we will stick to the low-speed model. The lift-to-drag ratio is

$$\frac{c_L}{c_{D_0} + k c_L^2} \tag{15.10}$$

which is a maximum when

$$c_L = \sqrt{\frac{c_{D_0}}{k}} \tag{15.11}$$

This is the lift coefficient at which we would like to operate in steady flight since it's where we get the most lift for the least drag.

The dynamic pressure, the pressure due to the motion of the aircraft, is

$$q = \frac{1}{2} \rho (u_r^2 + u_t^2) \tag{15.12}$$

where $\sqrt{u_r^2 + u_t^2}$ is the speed and ρ is the atmospheric density. The magnitudes of lift and drag are

$$L = q c_L s \tag{15.13}$$
$$D = q c_D s \tag{15.14}$$

where s is the wetted area. A wetted area is the area of the vehicle that contributes to lift and drag. The lift force is normal to the velocity vector, and the drag force is along the velocity vector. The unit velocity vector is

$$\vec{u} = \frac{\begin{bmatrix} u_r \\ u_t \end{bmatrix}}{\sqrt{u_r^2 + u_t^2}} \tag{15.15}$$

We use the arrow, $\vec{}$, to denote a vector, which in MATLAB is a multielement array. The drag vector is

$$\vec{D} = D\vec{u} \tag{15.16}$$

The lift vector is a 90-degree rotation and is

$$\vec{L} = L \begin{bmatrix} 0 & 1 \\ -1 & 0 \end{bmatrix} \vec{u} \tag{15.17}$$

In both cases, the radial component is the top element, and the bottom is the tangential component. The thrust vector is

$$\vec{T} = T \begin{bmatrix} \sin\alpha \\ \cos\alpha \end{bmatrix} \tag{15.18}$$

We start in a circular orbit. To begin reentry, we increase the angle of attack to increase the drag. Remember, we are using a simple aerodynamics model. This causes the ramjet to descend. We then adjust the angle of attack and thrust to reach our desired landing point. In this chapter, we will design one trajectory using optimal control, and then we will use reinforcement learning to design a trajectory. Our objective will be to reach the Titan surface with zero velocity.

15.3 Modeling the Titan Atmosphere

15.3.1 Problem

We want to compute the atmospheric density of Titan.

15.3.2 Solution

Use a database of Titan densities and altitudes and interpolate.

15.3.3 How It Works

The function `TitanAtmosphere` computes the Titan atmospheric density and temperature.

TitanAtmosphere.m

```
21
22  % Demo
23  if( nargin < 1 )
24    TitanAtmosphere(linspace(0,1400));
25    return
26  end
27
28  % Altitude
29  hD    = [0:10:100 150:50:900]; % km
30
31  % Density
32  rhoD  = [ 5.270e0  3.467e0  2.144e0  1.233e0  6.731e-1 3.575e-1...
33            1.825e-1 8.264e-2 4.788e-2 3.155e-2 2.219e-2 5.024e-3...
34            1.393e-3 4.612e-4 1.673e-4 6.282e-5 2.438e-5 9.806e-6...
35            4.114e-6 1.718e-6 7.129e-7 2.931e-7 1.173e-7 4.653e-8...
36            1.868e-8 7.934e-9 3.404e-9];
37
38  % Temperature
39  tD    = [  92.89   83.29   76.44   72.20   70.51   71.16   76.62 103.46...
40           122.88 133.97 140.80 159.23 173.76 181.72 181.72 181.72...
41           181.72 180.81 173.96 167.06 160.09 153.04 148.62 148.62...
42           148.62 148.62 148.62];
43
44  % Speed of sound
45  aD    = [ 195.6 185.7 177.7 173.1 171.5 172.5 179.2 208.4 227.1...
46           237.1 243.1 258.6 270.1 276.2 276.2 276.2 276.2 275.5...
47           270.2 264.8 259.3 253.5 249.8 249.8 249.8 249.8 249.8];
```

The altitude is computed through linear interpolation. This means that intermediate values will be returned. If no outputs are requested, it produces a double-y plot with this code:

TitanAtmosphere.m

```
49  rho   = interp1(hD,rhoD,h,'linear'); % Density
50  t     = interp1(hD,tD,  h,'linear'); % Temperature
51  a     = interp1(hD,aD,  h,'linear'); % Speed of sound
```

The commands yyaxis right and yyaxis left set the axis handles to match the left and right axes with the plot. Typing TitanAtmosphere will result in two plots. The demo plots are shown in Figure 15.3.

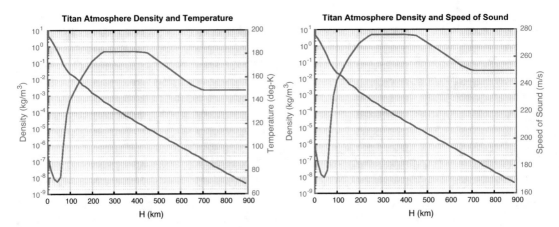

Figure 15.3: *Titan and its atmospheric density. The speed of sound and temperature curves have the same shape.*

15.4   Simulating the Aircraft

15.4.1 Problem

We want to numerically integrate the trajectory given aircraft parameters and the controls.

15.4.2 Solution

Write a function with a built-in integrator to integrate the equations of motion. Write a second function that returns the state derivative.

15.4.3 How It Works

The function RHS2DTitan models the two-dimensional dynamics of the vehicle. It includes gravity and aerodynamics. It is specialized for Titan.

RHS2DTitan.m

```
26  function [xDot, q] = RHS2DTitan( ~, x, d )
27
28  % Constants
29  rTitan = 2575;
30  mu     = 9.142117352579678e+03;
31  kMToM  = 1000;
32  nToKN  = 0.001;
33
34  % Return the default data structure
35  if( nargin < 1 )
36    xDot = DefaultDataStructure;
37    return
38  end
39
40  % To clarify the code use local variables
```

```
41   r         = x(1);
42   uR        = x(2);
43   uT        = x(3);
44
45   % Atmospheric density
46   h         = r - rTitan;
47   if( h <= 0 )
48     h = 0;
49   end
50   [rho,~] = TitanAtmosphere(h);
51
52   % Forces
53   uRM       = uR*kMToM;
54   uTM       = uT*kMToM;
55   w         = uRM^2 + uTM^2;
56
57   % Lift and drag
58   cL        = d.cLAlpha*d.alpha;
59   eps       = 0.8; % Oswald efficiency
60   k         = 1/(pi*d.aR*eps);
61   cD        = d.cD0 + k*cL^2;
62
63   q         = 0.5*rho*w;
64
65   % Direction of the velocity vector
66   u         = [uRM;uTM]/sqrt(w);
67   drag      = -cD*q*d.s*u;
68   lift      =  cL*q*d.s*[0 1;-1 0]*u;
69   c         = cos(d.alpha);
70   s         = sin(d.alpha);
71   thrust    = d.thrust*[c -s;s c]*u;
72   a         = nToKN*(thrust + drag + lift)/d.mass;
73
74   % State derivatives
75   uRDot     =  uT^2/r - mu/r^2 + a(1);
76   uTDot     = -uT*uR/r           + a(2);
77   xDot      = [uR;uRDot;uTDot];
```

Constants are at the beginning of the function. The next block of code returns the default data structure. Forces are converted to kN since the units of the states are km. This is followed by the dynamics code. The subfunction returns the default data structure. If you run the function, it will print the default data structure in the command window.

RHS2DTitan.m

```
80   function d = DefaultDataStructure
81
82   mu        = 9.142117352579678e+03;
83   r     = 2875;
84   d = struct('aR',1.7,'eps',0.9,'s',10,'cD0',0.006,'cLAlpha',2*pi,...
85     'mass',2000,'alpha',0,'thrust',0,'x0',[r;0;sqrt(mu/r)]);
86   d.states = {'r (km)' 'u_r (km/s)', 'u_t (km/s)'};
```

Most of the data structure is in the `struct` statement. The cell array is added afterward. If it was added in the `struct`, MATLAB would make a three-element structure, one for each cell array element. The default data structure gives the user an initial state and suggested state names for plotting. The default data structure puts the vehicle in a circular orbit.

A second function, `Simulation2DTitan`, runs the simulation. Its inputs are the same as `RHS2DTitan` with the addition of the control input, which is a two-by-n array of the angle of attacks and thrusts.

Simulation2DTitan.m

```
18
19  if ( nargin < 1 )
20    Demo
21    return
22  end
23
24  rTitan    = 2575;
25
26  n         = length(t);
27  xP        = zeros(3,n);
28  xP(:,1) = x;
29
30  for k = 2:n
31    d.alpha   = u(1,k-1);
32    d.thrust = u(2,k-1);
33          rHS          = @(t,x,d) RHS2DTitan(x,t,d);
34    [~, x]     = ode113(rHS, [0, t(k)-t(k-1)], x, [], d );
35    x          = x(end,:)';
36
37    xP(:,k)   = x;
38    if ( x(1) - rTitan <= 0 )
39       break;
40    end
41  end
42
43  xP = xP(:,1:k);
44  t  = t(:,1:k);
45
46  if ( nargout < 1 )
47    [t,tL] = TimeLabel(t);
48    PlotSet(t,xP,'x label',tL,'y label',d.states,'figure title','Titan
          Simulation');
49    clear x
50  end
```

We use ode113 internally to give us flexibility in the size of the control time steps. ode113 ensures that we will keep the integration errors below certain bounds. We are using the built-in bounds in this case. The code is

Simulation2DTitan.m

```
33              rHS        = @(t,x,d) RHS2DTitan(x,t,d);
34    [~, x]    = ode113(rHS, [0, t(k)-t(k-1)], x, [], d );
```

The first line rearranges the order of inputs of RHS2DTitan to be compatible with ode113. ode113 returns all of the intermediate values in an n-by-three array so we just grab the last set.

The built-in demo starts in a circular orbit with a high angle of attack which starts reentry.

Simulation2DTitan.m

```
52    %% Simulation2DTitan>Demo
53    function Demo
54
55    d = RHS2DTitan;
56    t = linspace(0,120000,1000);
57    u = zeros(2,1000);
58    u = [0.0001*ones(1,1000);
59         0.00*ones(1,1000)];
60    Simulation2DTitan( d.x0, t, d, u );
```

Figure 15.4 shows the results of the demo. As expected, the aircraft begins reentry.

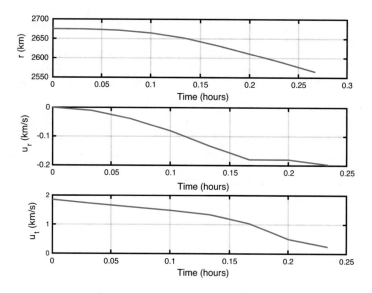

Figure 15.4: *Simulation using the built-in demo.*

15.5   Simulating Level Flight

15.5.1 Problem

We want to fly our aircraft level to the Titan surface.

15.5.2 Solution

Write a script calling the `Simulation2DTitan` function with the aircraft at the optimal angle of attack and sufficient thrust to maintain level flight.

15.5.3 How It Works

For level flight, we will operate at the optimal angle of attack, given in Equation 15.11, with enough thrust to overcome drag. The velocity needs to be fast enough to generate sufficient lift to balance gravity at the desired altitude. Let's assume that we want to fly at 1 km altitude above the surface of Titan. The script `LevelFlight.m` sets up and runs the simulation. We need equilibrium thrust, velocity, and the angle of attack. The easiest way to do this is with a search where the cost is the magnitude of the acceleration vector.

We first set up the simulation.

LevelFlight.m

```
1   %% Simulate level flight above Titan
2
3   %% Constants
4   rTitan     = 2575;
5   mu         = 9.142117352579678e+03;
6   kMToM      = 1000;
7
8   %% Get the default data structure
9   d          = RHS2DAero;
10
11  %% Compute the flight conditions and the controls
12
13  % Aircraft and flight parameters
14  d.mass     = 2000; % kg
15  d.s        = 20; % m^2
16  altitude   = 100; % km
```

We then use `fminsearch` to find the angle of attack, thrust, and velocity that make the accelerations as small as possible.

LevelFlight.m

```
18   % Find the equilibrium velocity and control
19   c          = [400;0.06;0]; % Initial control [thrust;v;alpha]
20   r          = d.rPlanet + altitude;
21
22   % Use a numerical search
23   Options    = optimset;
24   fun        = @(c) Cost(c,d,r);
25   c          = fminsearch( fun, c, Options );
```

The cost function is

LevelFlight.m

```
45   function c = Cost( u, d, r )
46
47   x(1)       = r;
48   d.thrust   = u(1);
49   x(3)       = u(2);
50   d.alpha    = u(3);
51
52   xDot       = RHS2DAero(x,0,d);
53
54   c          = Mag(xDot);
55
56   end
```

It calls `RHS2DTitan` and computes the state derivative. In equilibrium flight, this should be zero. It computes the magnitude of the acceleration vector and uses this as the cost. We then run the simulation.

LevelFlight.m

```
30   %% Run the simulation
31   n          = 2000;
32   t          = linspace(0,3600,n);
33   x          = [r;0;c(2)];
34   u          = [d.alpha*ones(1,n);d.thrust*ones(1,n)];
35   Simulation2DAero( x, t, d, u );
36
37   %% Print out the parameters
38   fprintf('Angle of attack          %8.2f deg\n',d.alpha*180/pi);
39   fprintf('Velocity                 %8.2f m/s\n',c(2)*kMToM);
40   fprintf('Thrust                   %8.2f N\n',d.thrust);
41   fprintf('Mass                     %8.2f kg\n',d.mass);
42   fprintf('Wetted area              %8.2f m^2\n',d.s);
```

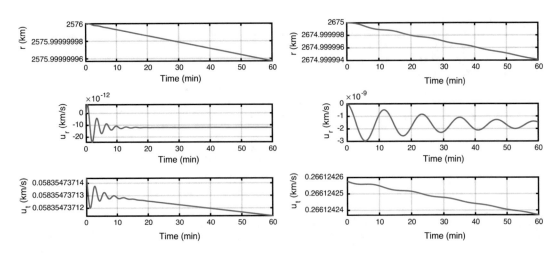

Figure 15.5: *Level flight simulations. The results are not perfect due to numerical errors. 1 km and 100 km altitude results are shown.*

The following prints out in the command window. It shows important parameters, states, and controls for 1 km altitude.

```
>> LevelFlight
Angle of attack            0.43 deg
Velocity                  58.35 m/s
Thrust                   373.95 N
Mass                    2000.00 kg
Wetted area               20.00 m^2
```

We also ran it at 100 km. The angle of attack is lower and the thrust is higher.

```
>> LevelFlight
Angle of attack            1.42 deg
Velocity                 266.12 m/s
Thrust                   177.26 N
Mass                    2000.00 kg
Wetted area               20.00 m^2
```

Figure 15.5 shows the results for both cases. The vehicle maintains level flight.

15.6   Optimal Trajectory

15.6.1 Problem

We want to produce an optimal trajectory to go from a starting point in an orbit to the surface.

15.6.2 Solution

Write a script using `fmincon` to produce an optimal trajectory.

15.6.3 How It Works

We will solve the constrained optimal control problem. The constraint is the end state, which must be

$$x = \begin{bmatrix} r_{\text{Titan}} \\ 0 \\ 0 \end{bmatrix} \tag{15.19}$$

We have three different cost calculations that could be minimized. `fmincon` tries to minimize the costs. The first is dynamic pressure:

$$q = \frac{1}{2}\rho v^2 \tag{15.20}$$

This is the aerodynamic load on the vehicle. Stagnation temperature is the second:

$$T_0 = T_a\left(1 + \frac{1}{2}f(\gamma - 1)M^2\right) \tag{15.21}$$

where T_a is the ambient temperature, γ is the ratio of specific heats, and M is the Mach number, which is the speed divided by the speed of sound. The last is the heating rate [45]:

$$r \approx \sqrt{\rho}v^3 \tag{15.22}$$

In all cases, the cost will be the mean of these quantities over the trajectory.

The first step is to find a trajectory that reaches the surface. We write a script `Landing.m` for that purpose.

Landing.m

```
1   %% Simulate level flight above Titan
2
3   %% Constants
4   rTitan        = 2575;
5   mu            = 9.142117352579678e+03;
6   kMToM         = 1000;
7
8   %% Get the default data structure
9   d             = RHS2DTitan;
10
11  %% Compute the flight conditions and the controls
12  d.mass        = 2000; % kg
13  d.s           = 20; % m^2
14  altitude      = 100; % km
15  r             = rTitan + altitude;
16  d.thrust      = 0;
17  d.alpha       = 0.0;
```

```
18  tEnd           = 20*60;
19
20  %% Run the simulation
21  n               = 2000;
22  t               = linspace(0,tEnd,n);
23  x               = [r;0;sqrt(mu/r)];
24  u               = [d.alpha*ones(1,n);d.thrust*ones(1,n)];
25  [~,xP,t]        = Simulation2DTitan( x, t, d, u );
26  [t,tL]          = TimeLabel(t);
27  d.states{1}     = 'Altitude (km)';
28  xP(1,:)         = xP(1,:) - rTitan;
29  PlotSet(t,xP,'x label',tL,'y label',d.states,'figure title','Reentry')
```

We start in a circular orbit and set the angle of attack to zero and zero thrust. We hit the ground within 15 minutes with a high vertical velocity and some tangential velocity. Figure 15.6 shows the trajectory. It shows that a good initial guess is zero angle of attack, and 15-minute duration.

Optimization using fmincon requires a constraint function and a cost function. The cost will be the mean dynamic pressure given in Equation 15.20. The constraint will be the landing condition. Both require that we simulate start until ground contact. We don't care when ground contact occurs. We break up time into segments of equal length.

The controls will be piecewise continuous along the trajectory. The number of segments will impact the accuracy of the solution and the speed of the solution. More segments mean higher accuracy but can slow convergence.

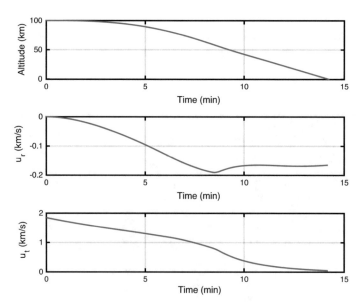

Figure 15.6: *Trajectory with no thrust and zero angle of attack.*

The constraint function is shown in the following code. It integrates the equations of motion until the vehicle hits the ground. Both functions use `Simulation2DTitan`.

TitanLandingConst2D.m

```
16  function [cIn, cEq] = TitanLandingConst2D( u, d )
17
18  rTitan  = 2575;
19  [~,xP]  = Simulation2DTitan( d.x, d.t, d, u );
20
21  % We don't have any nonlinear inequality contraints
22  cIn = [];
23
24  % Equality constraint
25  cEq = [xP(1,end)-rTitan;xP(2:3,end)];
```

One of three costs can be selected: mean stagnation temperature, mean dynamic pressure, and mean heating rate. They are computed by integrating the equations of motion over the trajectory.

TitanLandingCost2D.m

```
14  function cost = TitanLandingCost2D( u, d )
15
16  [~,xP]  = Simulation2DTitan( d.x, d.t, d, u );
17  cost    = CostCalculation( xP, d.costType );
18
19  %% Dynamic pressure
20  function cost = CostCalculation( x, type )
21
22  h           = x(1,:)- 2575; % Altitude (km)
23  j           = h<0;
24  h(j)        = 0;
25  [rho, t, a] = TitanAtmosphere(h);
26  v           = 1e3*sqrt(x(2,:).^2 + x(3,:).^2); % m/s
27
28  switch type
29    case 'stagnation temperature'
30      gamma     = 1.4;
31      m         = v./a;
32      r         = 0.88;
33      t0    = t.*(1 + 0.5*r*(gamma - 1).*m.^2);
34      cost      = mean(t0);
35    case 'heating rate' % p 238 Wiesel
36      cost  = mean(sqrt(rho).*v.^3);
37    case 'dynamic pressure'
38      q     = 0.5*rho.*v.^2;
39      cost  = mean(q);
40    otherwise
41      error('%s not available',type);
42  end
```

As you can see, the process is quite numerically intensive since we are integrating the equations of motion twice. The script to do the optimization is shown in the following code:

OptimalTitanLanding.m

```
1   %% Optimal Titan Landing
2
3   rTitan          = 2575;
4   mu              = 9.142117352579678e+03;
5   h               = 100;
6   dT              = 1;
7   n               = 50;
8
9   d               = RHS2DTitan; % The data structure
10  d.n             = n; % Number of decision variable increments
11  d.tEnd          = 20*60; % From the Landing script
12  d.t             = NonuniformSequence(d.tEnd,d.n,@ExponentialDT);
13  d.costType      = 'heating rate';
14
15  % Initial state
16  r               = rTitan + h;
17  x               = [r;0;sqrt(mu/r)];
18  d.x             = x;
19
20  % fmincon options
21  opts            = optimset( 'Display','iter-detailed',...
22                              'TolFun',1e-4,...
23                              'algorithm','interior-point',...
24                              'TolCon',1e-4,...
25                              'MaxFunEvals',15000);
26
27  % The cost is the time to reach the final state vector
28  costFun = @(x) TitanLandingCost2D(x,d);
29
30  % The numerical integration of the state is in the constraint function
31  constFun        = @(x) TitanLandingConst2D(x,d);
32
33  % The decision variables are angle of attack and thrust
34  u0         = 0.0001*ones(2,n);
35
36  % Lower and upper bounds for the decision variables
37  oN         = ones(n,1);
38  lB         = zeros(2,n);
39  uB         = [ (pi/12)*oN ;400*oN ];
40
41  % Find the optimal decision variable
42  u               = fmincon(costFun,u0,[],[],[],[],lB,uB,constFun,opts);
43
44  %% Run the simulation
45  x               = [r;0;sqrt(mu/r)];
46  [~,xP,t]        = Simulation2DTitan( x, d.t, d, u );
47  [t,tL]          = TimeLabel(t);
```

```
48   u           = u(:,1:length(t));
49   d.states{1} = 'Altitude (km)';
50   xP(1,:)     = xP(1,:) - rTitan;
51   yL          = [d.states(:)' {'\alpha (rad)'} {'T (N)'}];
52   PlotSet(t,[xP;u],'x label',tL,'y label',yL,'figure title','Optimal
         Landing')
```

You'll note that we use a non-uniform time distribution. This is because near the ground we expect that the controller will need to make decisions more quickly. The function has a default of a linearly decreasing step size.

NonuniformSequence.m

```
13   function t = NonuniformSequence(tEnd,n,dTFun)
14
15   if( nargin < 1 )
16      NonuniformSequence(60*20,50,@Exponential)
17      return
18   end
19
20   if( nargin < 3 )
21      dTFun = @Linear;
22   end
23
24   dT = dTFun(n);
25
26   % Scale
27   s  = tEnd/sum(dT);
28
29   t  = zeros(1,n);
30   for k = 2:n
31      t(k) = t(k-1) + s*dT(k-1);
32   end
33
34   if( nargout < 1 )
35      PlotSet(1:n,t,'x label','step','y label','Value','figure title','Non-
           uniform distribution')
36      clear t
37   end
38
39   %% NonuniformDistribution>Linear
40   function dT = Linear(n)
41
42   dT = linspace(1,0.1,n-1);
```

The sequence is shown in Figure 15.7. Other sequences are possible. We use an exponentially decreasing step size.

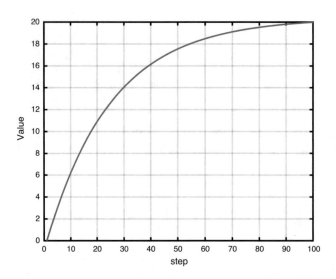

Figure 15.7: *Exponentially distributed step.*

ExponentialDT.m

```
13   function dT = ExponentialDT(n,f)
14
15   if( nargin < 1 )
16     NonuniformSequence(20,100,@ExponentialDT);
17     return
18   end
19
20   if( nargin < 2 )
21     f = 4;
22   end
23
24   h   = linspace(0,f,n);
25   dT  = exp(-h);
```

When you run the landing script, you will get the following output in the command line. The first column is the number of iterations. f(x) is the cost, in this case, the heating rate. The second column is the total function count. The fourth column, Feasibility, is how well it is matching the landing constraint of zero altitude and velocities. Ideally, it is zero when the constraint is matched. The first-order optimality is how close the solution is to the optimal solution. The norm step size is the norm (essentially magnitude) of the control step it is taking.

```
>> OptimalTitanLanding
```

Iter	F-count	f(x)	Feasibility	First-order optimality	Norm of step
0	101	1.832880e+08	1.273e+01	2.429e+10	
1	203	1.812246e+08	1.294e+01	4.069e+10	1.402e-04
2	304	1.799248e+08	1.289e+01	5.163e+10	2.207e-04
3	405	1.784495e+08	1.293e+01	4.786e+10	1.534e-03

4	511	1.754317e+08	2.184e+01	4.101e+10	3.541e-04
5	612	1.726764e+08	2.254e+01	4.224e+10	1.199e-02
6	713	1.726620e+08	2.254e+01	4.244e+10	9.581e-04
7	814	1.772045e+08	2.420e+00	2.339e+10	9.873e-04
8	921	1.782458e+08	1.265e+00	3.826e+10	4.015e-04
9	1022	1.782643e+08	1.264e+00	3.721e+10	2.822e-03
10	1125	1.783679e+08	1.242e+00	5.748e+10	7.499e-04

The end is shown as follows:

Iter	F-count	f(x)	Feasibility	First-order optimality	Norm of step
31	3306	1.037209e+08	3.961e-02	1.376e+09	1.718e-04
32	3413	1.037058e+08	3.959e-02	9.068e+09	7.366e-07
33	3522	1.037059e+08	3.959e-02	1.063e+10	6.471e-08
34	3625	1.037059e+08	3.959e-02	3.248e+09	1.439e-08
35	3726	1.037059e+08	3.959e-02	3.248e+09	1.473e-08
36	3827	1.037059e+08	3.959e-02	3.248e+09	1.450e-08
37	3928	1.037059e+08	3.959e-02	1.994e+09	1.444e-08

Figure 15.8 shows the optimal solution. It varies the angle of attack near the end. The mean heating rate drops by a third, and the constraints are met.

The solution shows that we should make the time steps even smaller near the end.

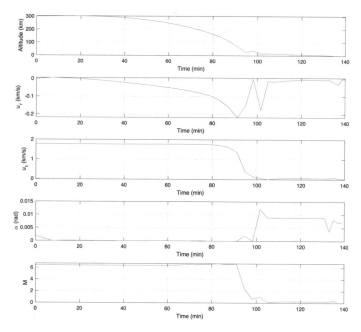

Figure 15.8: *Optimal landing.*

15.7   Reinforcement Example

15.7.1 Problem

We want to produce a trajectory to go from a starting point in an orbit to the surface.

15.7.2 Solution

Use reinforcement learning to produce the angle of attack to minimize the mean dynamic pressure while reaching zero altitudes at zero velocities.

15.7.3 How It Works

Figure 15.9 shows the pattern for reinforcement learning. It is essentially a trial and error approach. The reinforcement learning algorithm implements a policy that it updates using experimentation. In our problem, the goal is a policy of the angle of attack that allows the lander to reach the surface with zero velocity. The model is the same as was used in the optimization approach in the previous section. As with the optimization approach, no apriori algorithm is used. The reinforcement learning algorithm creates its algorithm from multiple attempts.

We will implement a Deep Deterministic Policy Gradient (DDPG) algorithm [21]. It is a model-free, online, off-policy reinforcement learning method. It works with continuous observations and actions. Off-policy agents create a buffer of observations and use that to update the policy. A DDPG agent is an actor-critic reinforcement learning agent that searches for an optimal policy.

The first step is to make a class to encapsulate the environment. The properties are constants. Calculations are not allowed here. The reward scale is the scaling of the position and velocity errors.

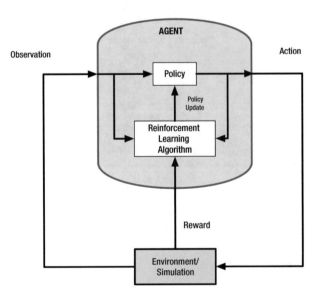

Figure 15.9: *Reinforcement learning.*

Properties for TitanLanderClass

```
1   classdef TitanLanderClass < rl.env.MATLABEnvironment
2       %TITANLANDERCLASS lander with 3 DOF class
3
4       %% Properties (set properties' attributes accordingly)
5       properties
6           % Specify and initialize environment's necessary properties
7           % Acceleration due to gravity in m/s^2
8
9           % Sample time
10          Ts       = 10;
11
12          % Control
13          Control = 0;
14
15          % Target
16          StateTarget = [2575;0;0];
17
18          % Initial State
19          StateInitial = [2875;0;sqrt(9.142117352579678e+03/2875)];
20      end
21
22      properties
23          % Initialize system State [r;v_r;v_t]
24          State = [2875;0;sqrt(9.142117352579678e+03/2875)];
25      end
26
27      properties(Access = protected)
28          % Initialize internal flag to indicate episode termination
29          IsDone = false
30      end
```

The next part is the necessary methods. This includes the class constructor, the `step` method, and the code `InitialObservation` method. The `step` method performs one integration time step.

Required methods for TitanLanderClass

```
33      %% Necessary Methods
34      methods
35          % Contructor method creates an instance of the environment
36          % Change class name and constructor name accordingly
37          function this = TitanLanderClass()
38              % Initialize Observation settings
39              ObservationInfo = rlNumericSpec([3 1]);
40              ObservationInfo.Name = 'Lander States';
41              ObservationInfo.Description = 'r v_r v_t';
42
43              % Initialize Action settings
44              ActionInfo        = rlNumericSpec([1 1]);
45              ActionInfo.Name   = 'Control';
```

```
46          ActionInfo.LowerLimit = 0;
47          ActionInfo.UpperLimit = pi/12;
48
49          % The following line implements built-in functions of RL
                env
50          this = this@rl.env.MATLABEnvironment(ObservationInfo,
                ActionInfo);
51
52          % Initialize property values and pre-compute necessary
                values
53          updateActionInfo(this);
54      end
55
56      % Apply system dynamics and simulates the environment with the
57      % given action for one step.
58      function [Observation,Reward,IsDone,LoggedSignals] = step(this,
            Action)
59          LoggedSignals = [];
60
61          % Get action
62          dRHS         = RHS2DTitan;
63          u            = getControl(this,Action);
64          dRHS.alpha   = u(1);
65
66              rHS          = @(t,x,d) RHS2DTitan(t,x,d);
67          [~, x]       = ode113(rHS, [0 this.Ts], this.State, [], dRHS
                );
68          this.State   = x(end,:)';
69
70          % The observation is the State
71          Observation = this.State;
72
73          altitude    = this.State(1) - 2575;
74          IsDone      = altitude <= 0.01;
75          this.IsDone = IsDone;
76          this.Control = u;
77
78          Reward = newReward(this);
79
80          % Use notifyEnvUpdated to signal that the environment has
                been update
81          notifyEnvUpdated(this);
82      end
83
84      % Reset environment to initial State and output initial
            observation
85      function InitialObservation = reset(this)
86          InitialObservation = this.StateInitial;
87          this.State = InitialObservation;
88
89          % (optional) use notifyEnvUpdated to signal that the
```

```
90              % environment has been updated (e.g. to update
                    visualization)
91              notifyEnvUpdated(this);
92          end
93      end
```

The remainder of the class is methods to set class members and graphics members. An important part is a reward. The reward is an exponential function of the magnitude of the rate error when the lander is near the ground. Otherwise it is the heating rate.

Reward for TitanLanderClass

```
106     % Reward function
107         function Reward = newReward(this)
108             h       = this.State(1)  - 2575;
109             x       = this.State(2:3);
110
111             % Mean heating rate
112             if( h < 0 )
113                 h = 0;
114             end
115             rho     = TitanAtmosphere(h);
116             v       = 1e3*sqrt(x(1)^2 + x(2)^2); % m/s
117             dP      = mean(sqrt(rho)*v^3)/6e7;
118
119             % Only care about velocity at zero altitude
120             if( h <= 0.01 )
121                 m = Mag(x(1:2));
122                 c = 1000*exp(-13*m);
123             else
124                 c = 0;
125             end
126             Reward = -dP + c;
127         end
```

The training is done in the following script. The agent is created in the call to `rlDDPG Agent`. The agent includes both the actor and the critic which are created by `rlDDPGAgent`. The training sets the weights for the actor and critic.

Lander training

```
1   env     = TitanLanderClass;
2
3   obsInfo = getObservationInfo(env);
4   actInfo = getActionInfo(env);
5   disp(obsInfo);
6   disp(actInfo);
7
8   initOpts = rlAgentInitializationOptions('NumHiddenUnit',128);
9   agent    = rlDDPGAgent(obsInfo,actInfo,initOpts);
10  agent.AgentOptions.NoiseOptions.StandardDeviationDecayRate = 1e-5;
```

```
11  agent.AgentOptions.NoiseOptions.StandardDeviationMin = 0.001*agent.
        ActionInfo.UpperLimit;
12  actorNet  = getModel(getActor(agent));
13  criticNet = getModel(getCritic(agent));
14
15  NewFigure('Actor Network')
16  plot(layerGraph(actorNet))
17  NewFigure('Critic Network')
18  plot(layerGraph(criticNet))
19  disp('Critic Network')
20  disp(criticNet.Layers)
21  disp('Actor Network')
22  disp(actorNet.Layers)
23
24  doTraining  = true;
25  maxsteps    = 2520;
26  maxepisodes = 5000;
27
28  trainingOpts = rlTrainingOptions(...
29      'MaxEpisodes',maxepisodes,...
30      'MaxStepsPerEpisode',maxsteps,...
31      'Verbose',true,...
32      'Plots','training-progress',...
33      'StopTrainingCriteria','EpisodeReward',...
34      'StopTrainingValue',595);
35
36  if( doTraining )
37      % Train the agent
38      trainingStats = train(agent,env,trainingOpts);
39  end
40
41  %% Simulate
42  simOptions = rlSimulationOptions('MaxSteps', 2000);
43  experience = sim(env,agent, simOptions);
44  t = experience.Observation.LanderStates.Time';
45  x = experience.Observation.LanderStates.Data;
46  u = experience.Action.Control.Data;
47  u =[0 reshape(u,1,size(u,3))]; % Scale
48  x = reshape(x,3,size(x,3));
49  xP = [x;u];
50
51  yL = {'H' 'V_R' 'V_T' '\alpha'};
52  xP(1,:) = xP(1,:) - 2575;
53  [t,tL]  = TimeLabel(t*env.Ts);
54  PlotSet(t,xP,'x label',tL,'y label',yL,'figure title','Time history');
```

We print out the observation information. The observations are the lander state. We don't set any observation limits.

```
obsInfo =

  rlNumericSpec with properties:

      LowerLimit: -Inf
      UpperLimit: Inf
            Name: "Lander States"
     Description: "r v_r v_t"
       Dimension: [3 1]
        DataType: "double"
```

The observations aren't given any limits. The action information is listed next. The action is angle of attack.

```
actInfo =

  rlNumericSpec with properties:

      LowerLimit: [0 0]
      UpperLimit: [0.7854 400]
            Name: "Control"
     Description: [0x0 string]
       Dimension: [2 1]
        DataType: "double"
```

The limits on the angle of attack are 0 and 45 degrees.

The actor network is shown in Figure 15.10 and listed as follows. The features are the three states. There are nine layers. This is the default network. You can write any neural network you would like. The actor network takes the state observation and creates the action which is to vary the angle of attack.

```
Actor Network
  9x1 Layer array with layers:

     1   'input_1'             Feature Input       3 features
     2   'fc_1'                Fully Connected     128 fully connected layer
     3   'relu_body'           ReLU                ReLU
     4   'fc_body'             Fully Connected     128 fully connected layer
     5   'body_output'         ReLU                ReLU
     6   'output'              Fully Connected     2 fully connected layer
     7   'tanh'                Tanh                Hyperbolic tangent
     8   'scale'               ScalingLayer        Scaling layer
     9   'RepresentationLoss'  Regression Output   mean-squared-error
```

The first layer is the "feature" input, which is the three measurements. These are passed to a 128-neuron fully connected layer that uses ReLU, rectified linear units, as the activation function. This is followed by another fully connected layer with the same activation function. The output layer is a fully connected layer with two neurons that uses a hyperbolic tangent activation function. The output of the activation layer is the vector control of the angle of attack.

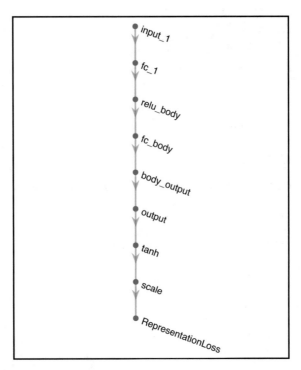

Figure 15.10: *Actor network. The input node is an array.*

The critic network shown in Figure 15.11 takes the states and the actions as inputs.

```
Critic  Network
  10x1 Layer array with layers:

       1    'concat'              Concatenation      Concatenation of 2 inputs along dimension 1
       2    'relu_body'           ReLU               ReLU
       3    'fc_body'             Fully Connected    128 fully connected layer
       4    'body_output'         ReLU               ReLU
       5    'input_1'             Feature Input      3 features
       6    'fc_1'                Fully Connected    128 fully connected layer
       7    'input_2'             Feature Input      2 features
       8    'fc_2'                Fully Connected    128 fully connected layer
       9    'output'              Fully Connected    1 fully connected layer
      10    'RepresentationLoss'  Regression Output  mean-squared-error
```

The training reports results in the command window. Each episode is 50 steps. It is breaking up the landing trajectory into 50 parts. The best reward is zero.

```
Episode:   31/5000 | Episode reward:   −173.49 | Episode steps:   820 | Average reward:   −51.65 | Step Count:
           32419 | Episode Q0:     16.95
Episode:   32/5000 | Episode reward:    192.06 | Episode steps:  1225 | Average reward:   −51.31 | Step Count:
           33644 | Episode Q0:     14.80
Episode:   33/5000 | Episode reward:    285.55 | Episode steps:  1172 | Average reward:    60.70 | Step Count:
           34816 | Episode Q0:     16.05
Episode:   34/5000 | Episode reward:   −191.11 | Episode steps:  1092 | Average reward:    60.05 | Step Count:
           35908 | Episode Q0:     15.98
Episode:   35/5000 | Episode reward:    −33.37 | Episode steps:  1332 | Average reward:    15.93 | Step Count:
           37240 | Episode Q0:     15.01
Episode:   36/5000 | Episode reward:   −185.27 | Episode steps:  1020 | Average reward:    13.57 | Step Count:
           38260 | Episode Q0:     15.41
Episode:   37/5000 | Episode reward:    −22.50 | Episode steps:  1404 | Average reward:   −29.34 | Step Count:
           39664 | Episode Q0:     11.70
Episode:   38/5000 | Episode reward:    688.72 | Episode steps:  1468 | Average reward:    51.29 | Step Count:
           41132 | Episode Q0:     11.13
```

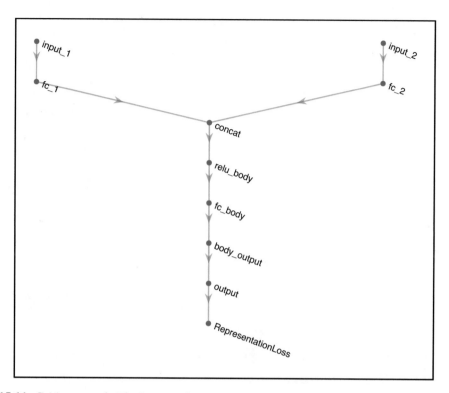

Figure 15.11: *Critic network. The input nodes are two arrays.*

Figure 15.12 shows the training window.

Episode Q0 is an estimate of the long-term reward at the start of each episode, using only the initial observation. The window also shows the reward for the current episode and the average reward. The highest reward, in this case, is zero.

You can halt training at any time, and the script will complete and run the simulation. The output of sim, experience, gives the simulation output. It is a fairly complex data structure.

```
>> experience

experience =

  struct with fields:

        Observation: [1x1 struct]
             Action: [1x1 struct]
             Reward: [1x1 timeseries]
             IsDone: [1x1 timeseries]
     SimulationInfo: [1x1 struct]
```

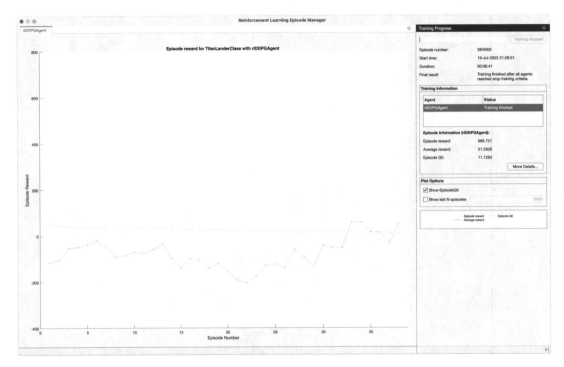

Figure 15.12: *Reinforcement learning training window. You can halt training at any tme.*

`experience.Observation.LanderStates` gives the time series for the simulation.

The landing is shown in Figure 15.13. The solution is different from the optimal solution. We do not give a negative reward for increasing the radius so it feels free to go to higher altitudes.

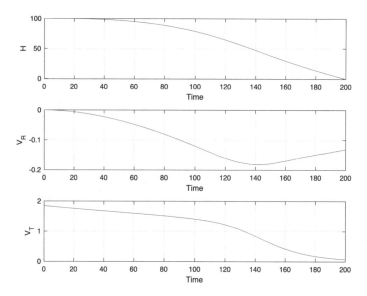

Figure 15.13: *Time history.*

Bibliography

[1] M.M.M. Al-Husari, B. Hendel, I.M. Jaimoukha, E.M. Kasenally, D.J.N. Limebeer, and A.Portone. Vertical stabilisation of Tokamak Plasmas. In *Proceedings of the 30th Conference on Decision and Control*, December 1992.

[2] Shaojie Bai, J. Zico Kolter, and Vladlen Koltun. An Empirical Evaluation of Generic Convolutional and Recurrent Networks for Sequence Modeling. *arXiv*, April 2018.

[3] D. Bertsekas. *Reinforcement Learning and Optimal Control*. Athena Scientific, 2019.

[4] Ilker Birbil and Shu-Chering Fang. An electromagnetism-like mechanism for global optimization. *Journal of Global Optimization*, 25:263–282, 03 2003.

[5] Léon Bottou, Frank E. Curtis, and Jorge Nocedal. Optimization methods for large-scale machine learning. *SIAM Review*, 60:223–311, 2016.

[6] A. Bryson and Y. Ho. *Applied Optimal Control*. Hemisphere Publishing Company, 1975.

[7] Barbara Cannas, Gabriele Murgia, A Fanni, Piergiorgio Sonato, Augusto Montisci, and M.K. Zedda. Dynamic Neural Networks for Prediction of Disruptions in Tokamaks. *CEUR Workshop Proceedings*, 284, 01 2007.

[8] Wroblewski D. and et al. Tokamak disruption alarm based on neural network model of high-beta limit. *Nuclear Fusion*, 37(725), 11 1997.

[9] Prafulla Dhariwal, Heewoo Jun, Christine Payne, Jong Wook Kim, Alec Radford, and Ilya Sutskever. Jukebox: A generative model for music, 2020.

[10] Dheeru Dua and Casey Graff. UCI machine learning repository, 2017.

[11] S. Dunbar. Stochastic Processes and Advanced Mathematical Finance Brief History of Mathematical Finance Rating Everyone. In *Semantic Scholar*, 2015.

[12] Pablo Ramon Escobal. *Methods of Orbit Determination*. Krieger Publishing Company, 1965.

[13] David Foster. *Generative Deep Learning*. O'Reilly Media, Inc., June 2019.

[14] David E. Goldberg. *Genetic Algorithms in Search, Optimization, and Machine Learning.* Addison-Wesley, 1988.

[15] S. Haykin. *Neural Networks.* Prentice-Hall, 1999.

[16] Cheng-Zhi Anna Huang, Ashish Vaswani, Jakob Uszkoreit, Noam Shazeer, Ian Simon, Curtis Hawthorne, Andrew M. Dai, Matthew D. Hoffman, Monica Dinculescu, and Douglas Eck. MUSIC TRANSFORMER:GENERATING MUSIC WITH LONG-TERM STRUCTURE. *arKiv*, 2019.

[17] Guang-Bin Huang, Qin-Yu Zhu, and Chee-Kheong Siew. Extreme learning machine: Theory and applications. *Neurocomputing*, 70(1):489–501, 2006. Neural Networks.

[18] P. Jackson. *Introduction to Expert Systems, Third Edition.* Addison-Wesley, 1999.

[19] Julian Kates-Harbeck, Alexey Svyatkovskiy, and William Tang. Predicting disruptive instabilities in controlled fusion plasmas through deep learning. *Nature*, 568:526–531, April 2019.

[20] Diederik P. Kingma and Jimmy Lei Ba. ADAM: A METHOD FOR STOCHASTIC OP-TIMIZATION. In *ICLR 2015*, 2015.

[21] Daniel S. Kolosa. A Reinforcement Learning Approach to Spacecraft Trajectory Optimization. Technical Report Dissertations 3542, Western Michigan University, 2019.

[22] Alex Krizhevsky, Ilya Sutskever, and Geoffrey E. Hinton. ImageNet Classification with Deep Convolutional Neural Networks. *Communications of the ACM*, 60(6), 2017.

[23] Y. Liang and JET EFDA Contributors. Overview of Edge Localized Modes Control in Tokamak Plasma. Technical Report Preprint of Paper for Fusion Science and Technology, JET-EFDA, 2017.

[24] Alan J Lockett and Risto Miikkulainen. Temporal Convolution Machines for Sequence Learning. Technical Report AI-09-04, Department of Computer Sciences, the University of Texas at Austin, 2009.

[25] Lopez. RNN, LSTM & GRU. *dProgrammer Lopez*, April 2019.

[26] Jere Schenck Meserole. *Detection Filters for Fault-Tolerant Control of Turbofan Engines.* Phd, Massachusetts Institute of Technology, 1981.

[27] Microsoft. sentence-completion. `https://drive.google.com/drive/folders/0B5eGOMdyHn2mWDYtQzlQeGNKa2s`, 2019.

[28] Aaron van den Oord, Sander Dieleman, Heiga Zen, Karen Simonyan, Oriol Vinyals, Alex Graves, Nal Kalchbrenner, Andrew Senior, and Koray Kavukcuoglu. Wavenet: A generative model for raw audio, 2016. cite arxiv:1609.03499.

[29] M. Paluszek, Y. Razin, G. Pajer, J. Mueller, and S.Thomas. *Spacecraft Attitude and Orbit Control: Third Edition*. Princeton Satellite Systems, 2019.

[30] Michael Phi. Illustrated Guide to LSTM's and GRU's: A step by step explanation. *Towards Data Science*, September 2018.

[31] G.A. Ratta, J..Vega, A. Murari, the EUROfusion MSTTeam, and JET Contributors. AUG-JET cross-tokamak disruption predictor. In *2nd IAEA TM*, 2017.

[32] L.M. Rasdi Rere, Mohamad Ivan Fanany, and Aniati MurniA rymurthy. Simulated annealing algorithm for deep learning. *Procedia Computer Science*, 72:137–144, 2015.

[33] Joseph Rocca. Understanding Variational Autoencoders (VAEs). *Towards Data Science*, September 2019.

[34] Elizabeth Rosenthal. Artificial Intelligence Approach Points to Bright Future for Fusion Energy. *Oak Ridge National Laboratory*, 2019.

[35] S. Russell and P. Norvig. *Artificial Intelligence A Modern Approach Third Edition*. Prentice-Hall, 2010.

[36] Paul A. Samuelson. Mathematics of speculative price. *SIAM Review*, 15(1):1–42, 1973.

[37] R.O. Sayer, Y.K.M. Peng, J.C. Wesley, S.C. Jardin, CA General Atomics, San Diego, and NJ Princeton Univ. ITER disruption modeling using TSC (Tokamak Simulation Code). Technical report, Oak Ridge National Laboratory, 11 1989.

[38] Luigi. Scibile. *Non-linear control of the plasma vertical position in a tokamak*. PhD thesis, University of Oxford, 1997.

[39] Richard Socher. *Recursive Deep Learning for Natural Language Processing and Computer Vision*. PhD thesis, Stanford University, August 2014.

[40] Russell Stewart. Maximum likelihood decoding with rnns - the good, the bad, and the ugly, 2016.

[41] Stephanie Thomas and Michael Paluszek. *MATLAB Machine Learning*. Apress, 2017.

[42] Stephanie Thomas and Michael Paluszek. *MATLAB Machine Learning Recipes: A Problem-Solution Approach*. Apress, 2019.

[43] P. Toiviainen and T. Eerola. MIDI toolbox 1.1. `https://github.com/miditoolbox/`, 2016.

[44] Phillip Wang, 2019.

[45] W. E. Wiesel. *Spaceflight Dynamics*. McGraw-Hill, 1988.

[46] Geoffrey Zweig and Chris J.C. Burges. The microsoft research sentence completion challenge. Technical Report MSR-TR-2011-129, Microsoft, December 2011.

Index

© Michael Paluszek, Stephanie Thomas, Eric Ham 2022
M. Paluszek et al., *Practical MATLAB Deep Learning*,
https://doi.org/10.1007/978-1-4842-7912-0

Printed in the United States
by Baker & Taylor Publisher Services